OPTICAL COMPONENTS FOR COMMUNICATIONS:
Principles and Applications

OPTICAL COMPONENTS FOR COMMUNICATIONS:
Principles and Applications

by

Ching-Fuh Lin
National Taiwan University, Taiwan, R.O.C.

KLUWER ACADEMIC PUBLISHERS
Boston / Dordrecht / New York / London

Distributors for North, Central and South America:
Kluwer Academic Publishers
101 Philip Drive
Assinippi Park
Norwell, Massachusetts 02061 USA
Telephone (781) 871-6600
Fax (781) 871-6528
E-Mail <kluwer@wkap.com>

Distributors for all other countries:
Kluwer Academic Publishers Group
Post Office Box 322
3300 AH Dordrecht, THE NETHERLANDS
Telephone 31 78 6576 000
Fax 31 78 6576 474
E-Mail <orderdept@wkap.nl>

 Electronic Services <http://www.wkap.nl>

Library of Congress Cataloging-in-Publication

Title: Optical Components for Commubications:
       Principles and Applications
Author (s): Ching-Fuh Lin

ISBN: 978-1-4419-5399-5

Printed in the United States of America

CONTENTS

PREFACE

PREFACE

Many people have used fiber-optic communication systems to transmit phone calls, emails, world-wide-web messages, and so on. However, not as many people understand the optics used in the fiber-optic communication systems. Therefore, the purpose of this book is to introduce "optics" for optical communications. As its name implies, this book focuses on the principles and applications of optical components for optical communications. This book tries to span a subject from the very abstract, fundamental principles of operation to the very specific real world applications of the technology. It is written in such a way that people with a background of general physics at the undergraduate level can easily understand the principles of optics and laser physics for optical communications. For those that are not familiar with the optical components used in fiber-optic communication systems, this book can act as an entry level guide.

Many different types of optical components are used in fiber-optic communication systems. Nonetheless, understanding their principles of operations is not as difficult as it appears to be because they are not completely dissimilar. One particular principle of optics may be applied to several types of components. On the other hand, several principles of optics may be used together for a single type of component. Therefore, it is better to discuss the individual optical principles before we study specific components. This book is thus organized in the following way. We described the general optics and physics related to optical communications first and then specific active components. Afterwards, we discuss individual optical principles for passive components, followed by discussion of specific passive components.

This book is not intended to describe any specific component to a level of detail required for expertise in that particular component or module. It does however provide a thorough understanding of a wide range of optical components. Thus people working on different component areas, module or system levels, or even in marketing can gain enough insight to make the best choice of components for their applications and know what tradeoffs to expect. It is also hoped that this book can simplify "communications" for people in the optical communication area by providing them the fundamental concepts of optics.

The use of optics in communications will continue to grow and eventually optical signals will reach most buildings and homes. People will then find it necessary to replace some optical components or modules at their homes or work places just like replacing their phone sets or phone cords. With an understanding of optical components, people can picture an optical-communication system with real devices and their actual structures instead of a pile of functional blocks, so they have confidence in using those components. At that time, optical components for communications will be as common as the optical disks that are sold on most street corners.

I sincerely thank Prof. Jingshown Wu, Prof. Way-Seen Wang, and Prof. Hen-Wai Tsao for their reviews and suggestions on the contents to be included in this book. I am particularly grateful to David Brooks for his corrections on the manuscripts. Without his help, reading this book will not be as pleasant as it is now. The collections of background information and references by Tien-Ting Hsiao, Yi-Shin Su, Chi-Chia Huang, and Chao-Hsin Wu are also highly appreciated.

Because the optical-communication systems and their optical components are still evolving, the present contents of this book can not cover all aspects of this field. In addition, no matter how careful I am in preparation of the manuscripts, it is not possible to make the book perfect. I believe that suggestions and criticism from the readers will provide the basis for a revised edition of this book and advance knowledge in this field, so I welcome hearing from readers interested in sharing their opinions.

Ching-Fuh Lin
Department of Electrical Engineering
National Taiwan University
Taipei, Taiwan

September, 2003

Chapter 1

INTRODUCTION TO OPTICAL COMMUNICATION

1. INTRODUCTION

In the 21st century, when people talk about optical communication, they are usually referring to optical fibers. In fact, optical communication has been used for thousands of years although optical physics were not understood in ancient times. In this chapter, we will first introduce the historical use of optical communication, followed by the description of emerging modern optical-fiber communication. To understand the significant progress of optical communication in the late 20th century, the fundamentals of light need to be understood. Therefore, we will discuss the details of optical physics in Section 1.3 and the light-matter interaction in Sections 1.4. The optical properties of transparent materials that are commonly encountered in optics will be introduced in Section 1.5, followed by the explanation of basic physics of optical fiber in Section 1.6. A brief description of modern optical communication system and optical components for communication is given in Section 1.7.

1.1 Brief history of optical communication

Human beings have used optical communication for more than 2000 years. In Europe, fire signals had been used to transmit messages around 800 BC. Two hundred years later, fire signals were transmitted via eight intermediate relay stations to pass final conquest of Troy to Argos, which is about 500 kilometers away. It was known as the torchpost of Agamemnon. Many other optical links using fire signals were later established. One very significant advancement was the coding-based optical communication system developed in Greek Polybios around 200 BC. The 24 Greek letters were coded with several torches. The system was run by trained operators. About 8 letters could be sent in one minute, corresponding to 0.67 bit/s. Although this transmission rate is much slower than modern optical-fiber communication, it already demonstrated the basic concept of communication.

In Asia, the application of optical communication also originated early. Optical communication was conducted by way of a beacon system. The first beacon in China probably appeared in the last years of Ying-Shang period (around 1300 B.C.) Later, in West Zhou dynasty, beacon communication became more developed and started to be applied in war and military defense (around 1100 B.C. – 771 B.C.). A beacon was lit to give alarm when the capital had been invaded. The dukes were required to come for help when the alarm was on. This system was eventually misused, causing the crash of a kingdom. It was recorded in *Shiji* [1] that King Zhouyou Wang played tricks on the dukes by lighting the beacon fire in order to delight his concubine Yuji. This careless trick lead to the perish of Western Zhou dynasty (around 779 B.C.). This event indicates the significance of optical communication even thousands of years ago.

According to ancient Chinese books, since fire could not be clearly seen in day time, wolf excrement was used as fuel to make smoke and thereby give alarm. That is why the beacon tower was also called wolf smoke tower. At night, firewood with sulfur was used instead.

The beacon system was highly developed in Tang Dynasty. (618-917 A.D.) It aggregated the beacon systems of previous dynasties and constructed a more perfect defense mechanism. Systematical operations, for example the beacon codes, were also thoroughly recorded in Song's book. [2] There were also detailed regulations about the tower locations, the form of tower construction, the delivery speed per day, and also the number of soldiers to guard each tower. Along the Great Wall, one tower was set up every 15 km. The system was regulated to delivery 1000 km per day. [3] Although the beacon system did not transmit as much and fast information as today's system, it showed the concept for maintaining and regulating communication in a systematic way.

Both fire and smoke signals are observed by human eyes, so the spectrum is in the optical frequency domain. Thus it is an optical-communication system. However, due to the lack of knowledge in optics, the optical spectrum was not well utilized in the ancient time. The optical signals generated by torch-coding or beacon system were switched by mechanical means, leading to the low transmission rate. In contrast, modern systems fully utilize the high frequency characteristics of optical signal. Techniques for high-speed switching to generate optical signals are highly developed, so the high bit rate operation of optical communication is possible.

2.   EMERGENCE OF MODERN OPTICAL-FIBER COMMUNICATION

Although optical communication had been used for thousands of years, the transmission rate remained below one bit per second even up to the 17th century. This was primarily due to the lack of knowledge in optics. In 1791, an optical semaphore telegraph was developed for communication. A star network using such an optical-relay semaphore telegraph was then established in France. In 1844, such a system included over 5000 kilometers of lines and afterwards extended to many countries in Europe and even in North Africa. However, after electricity was discovered, optical communication was abandoned and replaced by telecommunication using electronic signals. Later, in 1887, Hertz discovered the electromagnetic wave. Thereafter radio waves became the standard carrier for communication. In 1876, Alexander Graham Bell demonstrated the telephone and in 1895, Guglielmo Marchese Marconi developed radio for broadcast communication. Electronic signals and radio waves continued to play the main role for communication after World War II. Little attention was paid to optical communication until 1970. Nonetheless, the development of electronic communication systems enabled the transmission of huge amounts of information and created a demand for more information transmission than ever, leading to the rapid progress of optical-fiber communication since it was first proposed in 1966.

The progress in modern physics, particularly in lasers and optics, gave rise to the modern optical-communication system that operates at more than 10 Gbit/s, more than 10 orders of magnitudes faster than the primitive optical-communication systems that existed before 1600.

2.1   Advance of Optical Fiber

The guidance of light in glass, the material used to make optical fiber, was well known to glassblowers a long time ago. However, this phenomenon was not recorded until 1870 when England scientist John Tyndall described the guidance of light in a curved water stream. Later, theoretical analysis of electromagnetic waves in a cylindrical dielectric was carried out in early 20th century. Due to the huge signal loss in glass fibers of that era, using glass fiber for light transmission was not deemed practical. The signal loss at that time was in the thousands of dB/km, supporting only very short transmission distances.

In 1966, Kao and Hockam pointed out that the huge signal loss in glass fibers was not intrinsic, but was caused by impurities. As a result, many researchers started to purify glass to minimize the attenuation of light propagation in glass fibers. In 1970, Corning successfully fabricated glass fibers with signal loss of less than 20 dB/km. In 1972, the loss was further reduced to below 8 dB/km and reached the minimum. 0.2 dB/km for wavelengths at 1.55 μm in 1979. Fig. 1-1 shows the loss of optical fiber vs. wavelength. The range of loss less than 0.5 dB/km extends from 1.3 μm to 1.6 μm except at the absorption peak around 1.4 μm. In 2000, the OH absorption peak was even removed. Thus the low-loss window extends from 1.3 μm to 1.6 μm, giving a very broad bandwidth for optical-signal transmission.

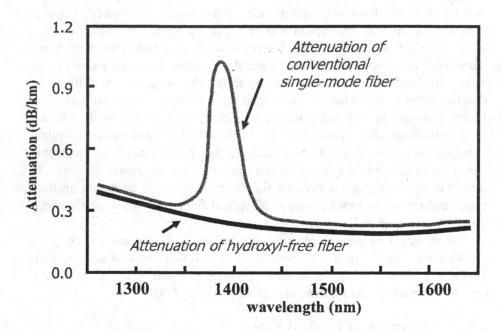

Figure 1-1. Variation of fiber loss with wavelength.

2.2 Advance of Laser Diodes

In addition to the progress of glass fiber, the laser technology for high-quality light generation also significantly advances the optical-fiber communication. The broad bandwidth of a low-loss window can only be realized when the light is modulated at very high frequency. The invention of lasers gave birth to the high coherent light that can be modulated at frequencies of tens of GHz. The progress of laser technology occurred concurrently with glass fiber technology. In 1960, the first laser was invented soon followed by the laser diode in 1962. Laser diodes (LDs) are particularly useful in optical-fiber communication. The continuous-wave (cw) operation of laser diodes at room temperature was first demonstrated in 1970. Laser diodes with lifetimes in excess of 100,000 hours were available in the market in 1979. Those breakthroughs enabled the first optical-fiber communication link to appear in 1983, operating at a wavelength of 820 nm. Higher Performance systems operating at 1310 nm were established in the United States, Europe, and Japan in 1984.

The tremendous advance of optical science and technology in the 20th century also greatly contributed to modern optical-fiber communications. For optical communications capable of carrying signals at THz, many issues of light propagation in optical fibers have to be taken care of. Those issues include:
– dispersion;
– chirp;
– good control of light reflection and/or transmission;
– precise separation of optical spectrum;
– high-speed modulation of light;
– efficient coupling of light into optical fibers.

These issues will be explained and discussed.

3 FUNDAMENTALS OF LIGHT

To understand optical communications, it is necessary to understand the basic characteristics of light. Fundamentally light behaves both as a wave and as a particle. Understanding both natures of light is necessary to understand its use in optical communications.

3.1 The Wave Nature of Light

When considering the wave nature of light, it acts as an electromagnetic wave that is composed of electric and magnetic fields. Both electric and magnetic fields oscillate perpendicularly to the direction that light is propagating. Fig. 1-2 is a snapshot of an electromagnetic wave. Assuming that the propagation is along the z-axis, the electric field oscillates along the x-axis, while the magnetic field oscillates along the y-axis. It is also possible that the electric field oscillates along the y-axis while the magnetic field oscillates along the x-axis.

The distance between the two neighboring peak positions of the oscillating electric field is called the wavelength, usually represented by the Greek letter λ. The distance between the two neighboring peak positions of the oscillating magnetic field is also equal to the wavelength. However, the peak position of the electric field is not the same as that of the magnetic field.

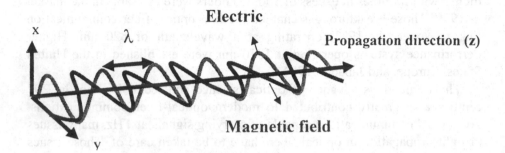

Figure 1-2. A schematic of snap-shot of electromagnetic wave.

For all kinds of waves, the speed of the wave is equal to the product of wavelength and frequency. An electromagnetic wave is a type of wave and so it's speed is given by the following formula.

$$\lambda \times \nu = c \tag{1-1}$$

For electromagnetic waves, the speed is designated by "c", commonly referred to as the speed of light, ν in this equation is the oscillating frequency of both the electric and magnetic fields.

A very important property for wave nature of light is the interference phenomenon. As shown in Fig.1-3 (a), when the electromagnetic wave propagates through two slits, interference patterns with bright and dark fringes will be seen on a screen placed at a distance behind the two slits. Another property of light's wave nature is diffraction. As illustrated in Fig. 1-3(b), the wave passing through a hole will expand to a region larger than the geometrically projected shape. In addition to the size increase, diffraction rings will also be produced, as shown in Fig. 1-3(b).

To be more specific, light is an electromagnetic wave with the electric field and the magnetic field following the Maxwell's equations. Maxwell's equations actually consist of four equations: Faraday's law, Ampere's law, Gauss's law for the electric field and Gauss's law for the magnetic field. They are given in the following formulae:

(a)

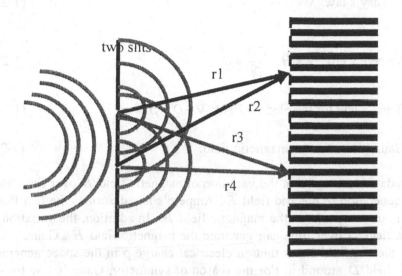

$|r1 - r2| = N\lambda$ Constructive interference: bright fringe

$|r3 - r4| = (N + 1/2)\lambda$ Destructive interference: dark fringe

(b)

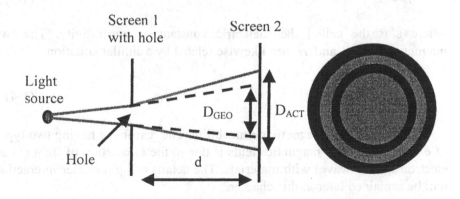

D_{GEO}: size of geometrically projected shape

D_{ACT}: size of actual shape due to diffraction

Figure 1-3. (a) Interference of wave propagating through two slits. (b) Diffraction of wave through a hole.

Faraday's law: $\nabla \times \bar{E} + \dfrac{\partial \bar{B}}{\partial t} = 0$ (1-2a)

Ampere's law: $\nabla \times \bar{H} - \dfrac{\partial \bar{D}}{\partial t} = \bar{J}$ (1-2b)

Gauss's law for the electric field: $\nabla \bullet \bar{D} = \rho$ (1-2c)

Gauss's law for the magnetic field: $\nabla \bullet \bar{B} = 0$ (1-2d)

Faraday's law says that the variation of magnetic field $\bar{B}$ over time leads to the generation of electric field $\bar{E}$. Ampere's law describes the fact that a flow of current results in the magnetic field $\bar{H}$. In addition, the variation of electric field $\bar{D}$ over time can generate the magnetic field $\bar{H}$. Gauss's law for the electric field states that an electrical charge ρ in the space generates electric field $\bar{D}$ around it. For the reason of symmetry, Gauss's law for the electric field is also written. Because there is no magnetic charge or monopole, the right hand side of Equation (1-2d) is equal to zero. This means that the existence of the magnetic field is not due to a charge, but due to the current flow or variation of electric field, as given by Ampere's law.

The two electric fields $\bar{D}$ and $\bar{E}$ are related by the following equation

$$\bar{D} = \varepsilon \bar{E} \tag{1-3}$$

where ε is the called the dielectric constant or permittivity. The two magnetic fields $\bar{B}$ and $\bar{H}$ are likewise related by a similar equation

$$\bar{B} = \mu \bar{H} \tag{1-4}$$

where μ is called magnetic permeability. The reason for having two types of electric fields and magnetic fields is due to the interaction of light (as an electromagnetic wave) with materials. The details of light-matter interaction will be explained later in this chapter.

In a vacuum, there is no interaction between light and material, so the relation between the two types of fields is very simple. The field $\bar{D}$ is proportional to the field $\bar{E}$ and the field $\bar{B}$ is also proportional to the field $\bar{H}$. They are related by the following equations.

$$\bar{D} = \varepsilon_0 \bar{E} \tag{1-5}$$

$$\bar{B} = \mu_0 \bar{H} \tag{1-6}$$

ε_0 and μ_0 in a vacuum are just constants with values given below.

$$\varepsilon_0 = \frac{1}{36\pi} \times 10^{-9} \frac{A \bullet s}{V \bullet m} \tag{1-7}$$

$$\mu_0 = 4\pi \times 10^{-7} \frac{V \bullet s}{A \bullet m} \tag{1-8}$$

Maxwell was the first person to put together the four equations: Faraday's law, Ampere's law, Gauss's law for the electric field, and Gauss's law for the magnetic field, so the four equations as a group are called Maxwell's equations. In addition, Maxwell discovered an extremely important result obtained from these four equations. With some mathematical derivation using vector analysis, the four equations lead to the following single equation for the electric field.

$$\nabla^2 \bar{E} - \mu\varepsilon \frac{\partial^2 \bar{E}}{\partial t^2} + (\nabla \log \mu) \times (\nabla \times \bar{E}) + \nabla(\bar{E} \bullet \nabla \log \varepsilon) = 0 \tag{1-9}$$

In a homogeneous medium, e.g. a vacuum, the dielectric constant ε and magnetic permeability μ do not vary in the space, so ($\nabla \log \mu$) and ($\nabla \log \varepsilon$) both equal zero. In this case the equation for the electric field becomes very simple.

$$\nabla^2 \bar{E} - \mu\varepsilon \frac{\partial^2 \bar{E}}{\partial t^2} = 0 \tag{1-10}$$

In fact, an equation similar to Eq. (1-10) can be obtained for the magnetic field as well. In a homogeneous medium, it can be reduced to the following simple equation.

$$\nabla^2 \bar{H} - \mu\varepsilon \frac{\partial^2 \bar{H}}{\partial t^2} = 0 \tag{1-11}$$

Equations (1-10) and (1-11) are actually the wave equations. In other words, the electromagnetic fields behave with the nature of a wave, so light exhibits a wave nature. From the above wave equation, we know that the propagation speed of wave is equal to $1/\sqrt{\mu\varepsilon}$. In a vacuum, it is equal to $1/\sqrt{\mu_0\varepsilon_0}$, which has a constant value of 3 x 10^{10} cm/s. When transmitted through materials, the propagation speed, $1/\sqrt{\mu\varepsilon}$, is less than $1/\sqrt{\mu_0\varepsilon_0}$. The ratio of light speed in vacuum to the light speed in the material is defined as the refractive index of this material.

n (refractive index) = light speed in vacuum / light speed in material

$$= \frac{\sqrt{\mu\varepsilon}}{\sqrt{\mu_0\varepsilon_0}} \qquad\qquad\qquad (1-12)$$

Table 1-1 provides the refractive indices of some materials.

Table 1-1 Refractive index of some materials

Air	n=1.000278
water	n=1.33
fused silica	n=1.46
crystal quartz	n=1.55
optical glass	n=1.51-1.81
sapphire	n=1.77
diamond	n=2.43

3.2   The Particle Nature of Light

In the early twentieth century, scientists such as Einstein and Compton discovered that light also has a particle nature. That is, light has behaviors similar to a particle. When behaving like a particle, a single particle of light is called a photon. A photon has an energy E equal to hν, where h is the Plank constant and ν is the frequency of light. Because the product of frequency and wavelength is equal to the speed of light, the energy of a photon is related to the wavelength by the following equation.

$$E = h\frac{c}{\lambda} \qquad (1\text{-}13)$$

Substituting the values of the Plank constant and the speed of light into the above equation, we obtain the following simple formula to calculate the photon energy directly from the wavelength of light.

$$E = \frac{1.24}{\lambda}(eV) \qquad (1\text{-}14)$$

The wavelength λ for the above formula is in the unit of μm. For example, if the wavelength is 1.55 μm, the photon energy is 0.8 eV. (1 eV = 1.6 x 10^{-19} joule) If the wavelength is 0.65μm, then the photon energy is 1.91 eV. It is very clear that the shorter the wavelength of the light, the larger the energy of the photon.

From the viewpoint of light's particle nature, the energy of light does not vary continuously. Instead, the energy is a multiple of a single photon. That is, $E = n\ h\nu$, where n is a positive integer. For example, assuming that the total energy of light is 10 eV, this light could consist of 10 photons. Each photon has 1 eV of energy. This light could also consist of a single photon with energy of 10 eV. However, it is not possible that this light consists of a half photon of 20 eV. In brief, the particle nature of light tells us that the appearance or disappearance of light is in the unit of a photon energy.

The particle point of view is particularly important for light absorption into or emission from materials. If the energy of light is absorbed by the material, this energy is always a multiple of photon energy ($h\nu$). Similarly, when materials emit light (like lasers or other light sources), the emitted energy is also a multiple of photon energy ($h\nu$).

4   INTERACTION OF LIGHT WITH MATTER

In this section, we will discuss the interaction of light with materials from the viewpoint of an electromagnetic wave. According to Maxwell's equations, electromagnetic fields interact with charge particles. Because the materials are composed of atoms, which consist of positively charged nuclei and negatively charged electrons, the electromagnetic wave certainly has interaction with materials. Such interactions can be described using the following three equations

$$\bar{J} = \sigma\bar{E} \qquad (1\text{-}15)$$

$$\bar{D} = \varepsilon \bar{E} = \varepsilon_0 \bar{E} + \bar{P} \qquad\qquad\qquad (1\text{-}16)$$

$$\bar{B} = \mu \bar{H} = \mu_0 \bar{H} + \bar{M} \qquad\qquad\qquad (1\text{-}17)$$

The above equations are called material equations or constitutive equations. They will be explained separately below.

4.1   Interaction of Electric Field with Conductors: $\bar{J} = \sigma \bar{E}$

In this equation, $\bar{J}$ represents current density and σ is the conductivity of the material. This equation describes the interaction of electric field with conductors. For conductors, σ is large. For insulators, σ is very small. In conductors, there are lots of carriers, which are charged particles like electrons, holes or ions. Carriers move in the existence of electric field, resulting in current. Therefore, as an electric field is applied to conductors, there is significant current flow in the conductor. In contrast, insulators have almost no carriers, so negligible current can be measured even with the existence of electric field. For simplicity, the insulator is usually treated as materials with $\sigma = 0$. As a result, the light-matter interaction of insulators is determined by Eqs. (1-16) and (1-17).

4.2   Interaction of Electric Field with Dielectrics: $\bar{D} = \varepsilon \bar{E} = \varepsilon_0 \bar{E} + \bar{P}$

In this equation, $\bar{P}$ is called polarization, which can be further described by

$$\bar{P} = \varepsilon_0 \chi_e \bar{E} . \qquad\qquad\qquad (1\text{-}18)$$

where χ_e is the electric susceptibility. Eq. (1-18) means that the externally applied electric field causes the polarization. Because the materials are composed of atoms, which consist of positively charged nuclei and negatively charged electrons, the applied electric field pulls positive and negative components toward opposite directions. Then as shown in Fig. 1-4, the positive nuclei and the center of negative charges are not at the same location. As a result, the local separation of positive and negative charges forms a dipole. Polarization is a collection of many such dipoles in the material. This dipole then again generates additional electric field. Thus $\bar{D}$

field is the sum of the externally applied field $\varepsilon_0 \bar{E}$ and the resulting polarization field $\bar{P}$.

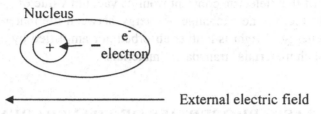

Figure 1-4. A schematic of an atom under the applied electric field. A dipole is formed.

Substituting Eq. (1-18) into Eq. (1-16), we obtain

$$\bar{D} = \varepsilon\bar{E} = \varepsilon_0\bar{E} + \varepsilon_0\chi_e\bar{E} = \varepsilon_0(1 + \chi_e)\bar{E} \tag{1-19}$$

Thus

$$\varepsilon = \varepsilon_0(1 + \chi_e) \tag{1-20}$$

4.3 Interaction of Electric Field with Magnetic Materials: $\bar{B} = \mu\bar{H} = \mu_0\bar{H} + \bar{M}$

Similar to the interaction of material with electric field, some materials interact with magnetic fields. A magnetic field can cause the magnetization of some materials, which is described by the symbol $\bar{M}$. $\bar{M}$ can be further described by $\bar{M} = \chi_m\bar{H}$, where χ_m is the magnetic susceptibility. If $\chi_m = 0$, the material is not magnetic, meaning that it can not interact with the magnetic field or its property cannot be changed by the applied magnetic field. If $\chi_m \neq 0$, then the material is magnetic and its property can be changed by the applied magnetic field.

4.4 Summary of Light-Matter Interaction

Most materials have $\chi_m = 0$, so magnetic fields have no influence on them. However, all materials have $\chi_e \neq 0$, so all materials have interaction with electric fields. Because an electromagnetic wave has both electric and magnetic fields, all materials will certainly interact with electromagnetic fields, primarily through the electric-field induced polarization effect.

According to the above discussion, we know that Eq. (1-16) is the most common situation for light-matter interaction. Such interaction will lead to the change of the dielectric constant from its vacuum value of ε_0 to ε. In such interactions, there is no exchange of energy between the material and light. When the energy of light is neither absorbed nor amplified by the materials, we called such materials "transparent materials".

5   BASIC PROPERTIES OF TRANSPARENT MATERIALS

Even without the exchange of energy between transparent material and light, there are still many important phenomena involved in the interaction between transparent materials and light. Those phenomena will be described in the following.

5.1   Reflection and Refraction

Reflection and refraction are common phenomena which occur at the boundary of optical components. When light propagates from one medium to another, some portion of light is reflected and some is transmitted. These phenomena can be explained using either ray optics or wave optics. Usually wave optics can explain not only the angles of reflection and refraction, but also the intensity ratio of reflection and transmission. Therefore, we will discuss the reflection and refraction from the viewpoint of wave optics.

5.1.1   Boundary Conditions for Electric and Magnetic Fields

Because light waves are electromagnetic waves, the boundary conditions for electric and magnetic fields are followed. Refer to Fig. 1-5, there is an interface between medium 1 and medium 2. At the interface boundary, the fields in the medium 1 and medium 2 are represented with subscripts 1 and 2, respectively. Starting from Maxwell's equations (Gauss's law for the electric field and Gauss's law for the magnetic field) and using vector analysis, we can obtain the following boundary conditions.

$$\hat{n} \bullet (\bar{B}_2 - \bar{B}_1) = 0 \qquad\qquad (1\text{-}21a)$$

$$\hat{n} \bullet (\bar{D}_2 - \bar{D}_1) = \rho_s \qquad\qquad (1\text{-}21b)$$

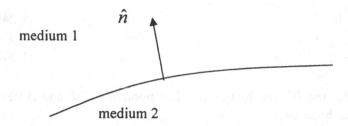

Figure 1-5. Interface of medium 1 and medium 2.

where $\hat{n}$ is the unit vector normal to the interface and ρ_s is the density of the surface charge (charge per unit area). For insulators, there is no surface charge, so $\rho_s = 0$. Then Eqs (1-21a) and (1-21b) become

$$\hat{n} \bullet (\bar{B}_2 - \bar{B}_1) = B_{2n} - B_{1n} = 0 \qquad (1\text{-}22a)$$

$$\hat{n} \bullet (\bar{D}_2 - \bar{D}_1) = D_{2n} - D_{1n} = 0 \qquad (1\text{-}22b)$$

where B_{2n}, B_{1n}, D_{2n}, and D_{1n} are the projections of fields along the direction normal to the boundary. Eqs. (1-22a) and (1-22b) give the results: $B_{2n} = B_{1n}$ and $D_{2n} = D_{1n}$. That is, the normal projections of B field and D field have to be continuous even at the interface of two media if medium 1 and medium 2 are both insulators.

At the interface, the tangential components E field and H field also have to be continuous, which can be derived as follows. Starting from Maxwell's equations (Faraday's law and Ampere's law) and using vector analysis, we can obtain the following boundary conditions.

$$\hat{n} \times (\bar{E}_2 - \bar{E}_1) = 0 \qquad (1\text{-}23a)$$

$$\hat{n} \times (\bar{H}_2 - \bar{H}_1) = \bar{K} \qquad (1\text{-}23b)$$

where $\bar{K}$ is the surface current density at the interface. Again, if both medium 1 and medium 2 are insulators, there is no surface current density, so $\bar{K} = 0$. Because the cross product of two vectors are perpendicular to the original two vectors, $\hat{n} \times \bar{E}_2$ points to the direction perpendicular to $\hat{n}$, which is the normal to the interface boundary. Thus $\hat{n} \times \bar{E}_2$ is the tangential direction along the interface. Therefore, for insulators, Eqs. (1-23a) and (1-23b) can be written as

$$E_{2t} - E_{1t} = 0 \tag{1-24a}$$

$$H_{2t} - H_{1t} = 0 \tag{1-24b}$$

where E_{2t}, E_{1t}, H_{2t}, and H_{1t} are the tangential components of E and H fields along the interface boundary.

5.1.2   Plane-Wave Solution of Wave Equation

Because the electromagnetic wave has to follow the wave equations (1-10) and (1-11), the mathematical function that describes waves should satisfy the wave equation. It is well known that many mathematical functions can satisfy the wave equation. One of the simplest forms is as follows

$$\varphi(\vec{r},t) = Ae^{j(\omega t - \vec{r} \bullet \vec{k})} \tag{1-25}$$

where $\vec{r}$ is the spatial coordinate and $\vec{k}$ is the wave vector with

$$|\vec{k}| \equiv k = \omega\sqrt{\mu\varepsilon} = \frac{n}{c}\omega \tag{1-26}$$

The wave with the mathematical form of (1-25) is called plane wave. A plane wave expressed by Eq. (1-25) has the following two basic characteristics.
1.   It has a unique direction of propagation which is the same as the wave vector $\vec{k}$.
2.   The positions of its constant phase form a plane. This constant phase plane is also called as the wave front. In other words, its phase front is a plane.

Due to the simple and clear characteristics of a plane wave, the reflection and transmission of a plane wave at the interface between two media is particularly simple. It gives the simplest and the most important concept for the understanding of light-matter interaction. The situation is illustrated in Fig. 1-6.

The boundary of medium 1 and medium 2 is on the x-y plane, so the normal of the boundary is parallel to the z-axis. Fig. 1-6 shows the x-axis and z-axis. The z-axis is assumed to be the normal of the boundary. A plane wave is incident on the boundary. Its direction of propagation, $\vec{k}_i$, is on the x-z plane. The incident angle, defined as the angle between the $\vec{k}_i$ vector and the normal of the boundary (z-axis), is θ_i. The electric field is perpendicular to the wave vector $\bar{k}_i$. This incident electric field is decomposed into two components, $\bar{E}_{//}$ and $\bar{E}_{\perp}$. $\bar{E}_{//}$ is the component on

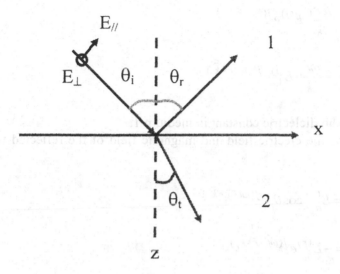

Figure 1-6. Reflection and transmission of a plane wave at a boundary located on the x-y plane.

the x-z plane, while $\bar{E}_{\perp}$ is the component parallel to the y-axis. Certainly, both $\bar{E}_{//}$ and $\bar{E}_{\perp}$ are perpendicular to the wave vector $\bar{k}_i$. This electric field can also be expressed using the Cartesian coordinates, i.e., $\bar{E} = \hat{x}E_x + \hat{y}E_y + \hat{z}E_z$ with

$$E_x^{(i)} = E_{//}^{(i)} \cos\theta_i e^{j(\omega t - \bar{r}\bullet\bar{k}_i)} \tag{1-27a}$$

$$E_y^{(i)} = -E_{\perp}^{(i)} e^{j(\omega t - \bar{r}\bullet\bar{k}_i)} \tag{1-27b}$$

$$E_z^{(i)} = -E_{//}^{(i)} \sin\theta_i e^{j(\omega t - \bar{r}\bullet\bar{k}_i)} \tag{1-27c}$$

where the superscript (i) indicates the incident field. According to Maxwell's Eqs., the magnetic field is related to the electric fields, so the x-, y-, and z-components of the magnetic field can be obtained.

$$H_x^{(i)} = \frac{\omega \varepsilon_1}{k} E_\perp^{(i)} \cos\theta_i e^{j(\omega t - \bar{r} \cdot \bar{k}_i)}$$

(1-28a)

$$H_y^{(i)} = \frac{\omega \varepsilon_1}{k} E_{//}^{(i)} e^{j(\omega t - \bar{r} \cdot \bar{k}_i)}$$

(1-28b)

$$H_z^{(i)} = -\frac{\omega \varepsilon_1}{k} E_\perp^{(i)} e^{j(\omega t - \bar{r} \cdot \bar{k}_i)}$$

(1-28c)

where ε_1 is the dielectric constant in medium 1.

Similarly, the electric field and magnetic field of the reflected wave are as follows.

$$E_x^{(r)} = E_{//}^{(r)} \cos\theta_r e^{j(\omega t - \bar{r} \cdot \bar{k}_r)}$$

(1-29a)

$$E_y^{(r)} = -E_\perp^{(r)} e^{j(\omega t - \bar{r} \cdot \bar{k}_r)}$$

(1-29b)

$$E_z^{(r)} = -E_{//}^{(r)} \sin\theta_r e^{j(\omega t - \bar{r} \cdot \bar{k}_r)}$$

(1-29c)

$$H_x^{(r)} = \frac{\omega \varepsilon_1}{k} E_\perp^{(r)} \cos\theta_i e^{j(\omega t - \bar{r} \cdot \bar{k}_r)}$$

(1-30a)

$$H_y^{(r)} = \frac{\omega \varepsilon_1}{k} E_{//}^{(r)} e^{j(\omega t - \bar{r} \cdot \bar{k}_r)}$$

(1-30b)

$$H_z^{(r)} = -\frac{\omega \varepsilon_1}{k} E_\perp^{(r)} e^{j(\omega t - \bar{r} \cdot \bar{k}_r)}$$

(1-30c)

where the superscript (r) indicates the reflected field. Also, the transmitted wave is as follows.

$$E_x^{(t)} = E_{//}^{(t)} \cos\theta_t e^{j(\omega t - \bar{r} \cdot \bar{k}_t)}$$

(1-31a)

$$E_y^{(t)} = -E_\perp^{(t)} e^{j(\omega t - \bar{r} \cdot \bar{k}_t)}$$

(1-31b)

$$E_z^{(t)} = -E_{//}^{(t)} \sin\theta_t e^{j(\omega t - \bar{r} \cdot \bar{k}_t)}$$

(1-31c)

$$H_x^{(t)} = \frac{\omega \varepsilon_2}{k} E_\perp^{(t)} \cos\theta_i e^{j(\omega t - \bar{r} \cdot \bar{k}_t)} \tag{1-32a}$$

$$H_y^{(t)} = \frac{\omega \varepsilon_2}{k} E_{//}^{(t)} e^{j(\omega t - \bar{r} \cdot \bar{k}_t)} \tag{1-32b}$$

$$H_z^{(t)} = -\frac{\omega \varepsilon_2}{k} E_\perp^{(t)} e^{j(\omega t - \bar{r} \cdot \bar{k}_t)} \tag{1-32c}$$

where the superscript (t) indicates the transmitted field and ε_2 is the dielectric constant of the medium 2.

According to the boundary conditions expressed by Eqs. (1-24a) and (1-24b), the tangential components of electric field E and magnetic field H should be continuous, so

$$E_x^{(i)} + E_x^{(r)} = E_x^{(t)} \tag{1-33a}$$

$$E_y^{(i)} + E_y^{(r)} = E_y^{(t)} \tag{1-33b}$$

$$H_x^{(i)} + H_x^{(r)} = H_x^{(t)} \tag{1-33c}$$

$$H_y^{(i)} + H_y^{(r)} = H_y^{(t)} \tag{1-33d}$$

Substituting the electric fields and magnetic fields expressed in Eqs. (1-27) – (1-32) into Eqs. (1-33), we obtain the following results:

1. $\bar{r} \cdot \bar{k}_i = \bar{r} \cdot \bar{k}_r \Rightarrow xk_{ix} + yk_{iy} + zk_{iz} = xk_{rx} + yk_{ry} + zk_{rz}$ (1-34a)

$\bar{r} \cdot \bar{k}_i = \bar{r} \cdot \bar{k}_t \Rightarrow xk_{ix} + yk_{iy} + zk_{iz} = xk_{tx} + yk_{ty} + zk_{tz}$ (1-34b)

2. $\cos\theta_i (E_{//}^{(i)} - E_{//}^{(r)}) = \cos\theta_t E_{//}^{(t)}$ (1-35a)

$\sqrt{\varepsilon_1}(E_{//}^{(i)} + E_{//}^{(r)}) = \sqrt{\varepsilon_2} E_{//}^{(t)}$ (1-35b)

$\sqrt{\varepsilon_1} \cos\theta_i (E_\perp^{(i)} - E_\perp^{(r)}) = \sqrt{\varepsilon_2} \cos\theta_t E_\perp^{(t)}$ (1-35c)

$$(E_\perp^{(i)} + E_\perp^{(r)}) = E_\perp^{(t)} \tag{1-35d}$$

5.1.3 Laws of Reflection and Refraction

Eq. (1-34a) characterizes the relation between the incident angle and the reflection angle. Eq. (1-34b) characterizes the relation between the incident angle and the transmitted angle. Because the wave vector is on the x-z plane, $k_{iy} = k_{ry} = k_{ty} = 0$. Also the boundary is at the x-y plane, so z=0. Thus Eqs. (1-34a) and (1-34b) reduce to $k_{ix} = k_{rx}$ and $k_{ix} = k_{tx}$, where $k_{ix} = k_i \sin \theta_i = \omega \sqrt{\mu \varepsilon_1} \sin \theta_i$, $k_{rx} = k_r \sin \theta_r = \omega \sqrt{\mu \varepsilon_1} \sin \theta_r$, and $k_{tx} = k_t \sin \theta_t = \omega \sqrt{\mu \varepsilon_2} \sin \theta_t$. Therefore, Eq. (1-34a) gives the law of reflection

$$\sin \theta_i = \sin \theta_r \tag{1-36}$$

Also, Eq. (1-34b) gives the law of refraction (transmission), $\sqrt{\varepsilon_1} \sin \theta_i = \sqrt{\varepsilon_2} \sin \theta_t$, or in a form more familiar to most of people

$$n_1 \sin \theta_i = n_2 \sin \theta_t \tag{1-37}$$

5.1.4 Reflection and Transmission Coefficients

Eqs. (1-35a)-(1-35d) give the relation among the amplitudes of incident wave, reflected wave, and transmitted wave. Please notice that the $\bar{E}_{//}$ and $\bar{E}_\perp$ components are not related to one another. Eqs. (1-35a) and (1-35b) relate the amplitudes of incident wave, reflected wave, and transmitted wave for the $\bar{E}_{//}$ component. This component is called transverse magnetic (TM) polarization because its magnetic field (along the y-axis shown in Fig. 1-6) is perpendicular to the x-z plane, which is the plane that contains the wave vector and the normal of the boundary. On the other hand, Eqs. (1-35c) and (1-35d) relate the amplitudes of incident wave, reflected wave, and transmitted wave for the $\bar{E}_\perp$ component. It is called transverse electric (TE) polarization because its electric field is perpendicular to the plane formed by the wave vector and the normal of the boundary. In some cases, the TM component is also called a p-wave and the TE component is called an s-wave. The TE polarization and TM polarization can be treated as two independent modes because they do not influence each other. This is why we decompose the electric field to two components, $\bar{E}_{//}$ and $\bar{E}_{\perp}$, as shown in Fig. 1-6.

With proper mathematical manipulation, Eqs. (1-35a)-(1-35d) lead to the following ratios

For TM polarization (p-wave)

$$R_{//} \equiv \frac{E_{//}^{(r)}}{E_{//}^{(i)}} = -\frac{n_2 \cos\theta_i - n_1 \cos\theta_t}{n_2 \cos\theta_i + n_1 \cos\theta_t} \tag{1-38a}$$

$$T_{//} \equiv \frac{E_{//}^{(t)}}{E_{//}^{(i)}} = \frac{2n_1 \cos\theta_i}{n_2 \cos\theta_i + n_1 \cos\theta_t} \tag{1-38b}$$

For TE polarization (s-wave)

$$R_{\perp} \equiv \frac{E_{\perp}^{(r)}}{E_{\perp}^{(i)}} = \frac{n_1 \cos\theta_i - n_2 \cos\theta_t}{n_1 \cos\theta_i + n_2 \cos\theta_t} \tag{1-39a}$$

$$T_{\perp} \equiv \frac{E_{\perp}^{(t)}}{E_{\perp}^{(i)}} = \frac{2n_1 \cos\theta_i}{n_1 \cos\theta_i + n_2 \cos\theta_t} \tag{1-39b}$$

$R_{//}$ and $R_{\perp}$ are reflection coefficients, while $T_{//}$ and $T_{\perp}$ are transmission coefficients. They are the ratios of amplitudes instead of intensity. For light waves, the intensity ratio (or power ratio) is more frequently used than the amplitude ratio. The intensity of light is related to the amplitude of the electric field as follows

$$I = \frac{1}{2}\sqrt{\frac{\varepsilon}{\mu}} |\bar{E}|^2 \tag{1-40}$$

5.1.5   **Reflectivity and Ratio of Transmitted Intensity**

In fact, it makes better sense to consider the portions of reflected intensity and transmitted intensity rather than the amplitudes from the energy point of view. The intensity reflectivity (or simply reflectivity) is given by

For TM polarization (p-wave)

$$\mathcal{R}_{//} = |R_{//}|^2 \equiv \left| \frac{E_{//}^{(r)}}{E_{//}^{(i)}} \right|^2 = \left| \frac{n_2 \cos\theta_i - n_1 \cos\theta_t}{n_2 \cos\theta_i + n_1 \cos\theta_t} \right|^2 \tag{1-41a}$$

For TE polarization (s-wave)

$$\mathcal{R}_{\perp} = |R_{\perp}|^2 \equiv \left| \frac{E_{\perp}^{(r)}}{E_{\perp}^{(i)}} \right|^2 = \left| \frac{n_1 \cos\theta_i - n_2 \cos\theta_t}{n_1 \cos\theta_i + n_2 \cos\theta_t} \right|^2 \tag{1-41b}$$

The ratio of transmitted intensity is given for a TM polarization (p-wave) by:

$$\mathcal{T}_{//} = \frac{n_2}{n_1} |T_{//}|^2 \equiv \frac{n_2}{n_1} \left| \frac{E_{//}^{(t)}}{E_{//}^{(i)}} \right|^2 = \frac{n_2}{n_1} \left| \frac{2 n_1 \cos\theta_i}{n_2 \cos\theta_i + n_1 \cos\theta_t} \right|^2 \tag{1-42a}$$

and for a TE polarization (s-wave) by:

$$\mathcal{T}_{\perp} = \frac{n_2}{n_1} |T_{\perp}|^2 \equiv \frac{n_2}{n_1} \left| \frac{E_{\perp}^{(t)}}{E_{\perp}^{(i)}} \right|^2 = \frac{n_2}{n_1} \left| \frac{2 n_1 \cos\theta_i}{n_1 \cos\theta_i + n_2 \cos\theta_t} \right|^2 \tag{1-42b}$$

Please note that the refractive indices have to be included in Eqs. (1-42a) and (1-42b) because the incident light and transmitted light are in different media. If both medium 1 and medium 2 have no absorption, the total power is conserved, so $\mathcal{R}_{//} + \mathcal{T}_{//} = 1$ and $\mathcal{R}_{\perp} + \mathcal{T}_{\perp} = 1$.

Fig. 1-7(a) shows an example of the intensity reflectivity vs. the incident angle. In this example, the light is incident from the air to the glass with the refractive index of 1.5. If the light is incident from the glass to the air, the situation is similar except that the incident angle is different, as shown in Fig. 1-7 (b).

5.1.6 Total Reflection

Fig. 1-7(a) shows that the reflection of intensity approaches 100% only when the incident angle is close to $90°$. On the other hand, Fig. 1-7(b) shows that the reflection of intensity approaches 100% as the incident angle is near $41.8°$ no matter if the light is TE or TM polarized. Then one question appears. What happens if the incident angle is larger than $41.8°$?

We can answer the above question from two viewpoints. First, from the law of refraction, $n_1 \sin\theta_i = n_2 \sin\theta_t$ (Eq. (1-37)), when light is incident from the medium of a smaller refractive index to the medium of a larger refractive index ($n_1 < n_2$), the incident angle is always larger than the refractive angle ($\theta_i > \theta_t$). Therefore, there is always some portion of reflected light and some portion of transmitted light. As the incident angle is $90°$, it means that the light is incident along the direction parallel to the boundary between the two media. Thus there is no transmission of light. In other words, the light propagates entirely in medium 1. However, as light is incident from the medium of a larger refractive index to the medium of a smaller refractive index ($n_1 > n_2$), then the incident angle is smaller than the refractive angle ($\theta_i < \theta_t$). As a result, it is possible that the refractive angle already approaches $90°$, but the incident angle is still much less than $90°$. The incident angle for which the refractive angle is equal to $90°$ is called critical angle, θ_c.

$$\sin\theta_c = \frac{n_2}{n_1}\sin 90^o = \frac{n_2}{n_1} \tag{1-43}$$

At this angle θ_c and beyond, there is no transmission of light. That is, the light is reflected completely, called total reflection. This phenomenon can be further understood from the second viewpoint. From Eq. (1-37), $\sin\theta_t = \frac{n_1}{n_2}\sin\theta_i$. For $\theta_i > \theta_c$, $\sin\theta_t = \frac{n_1}{n_2}\sin\theta_i > \frac{n_1}{n_2}\sin\theta_c$. Using Eq.(1-43), we obtain $\sin\theta_t > 1$. Because $\cos\theta_t = \sqrt{1 - \sin^2\theta_t}$, so $\cos\theta_t$ is a pure imaginary number. Then Eqs. (1-38a) and Eq. (1-39a) become the form of $\frac{X - jY}{X + JY}$, where X and Y are real numbers and j represents the imaginary number, $j^2 = 1$. Therefore, the absolute value of $R_{//}$ and $R_{\perp}$ is of the form $\frac{\sqrt{X^2 + Y^2}}{\sqrt{X^2 + Y^2}}$ (=1), which makes the intensity reflectivity ($\mathcal{R}_{//}$ and $\mathcal{R}_{\perp}$) equal to one. It clearly explains that the incident light is reflected completely, so there is no transmission of light. The total reflection only occurs for light

(a)

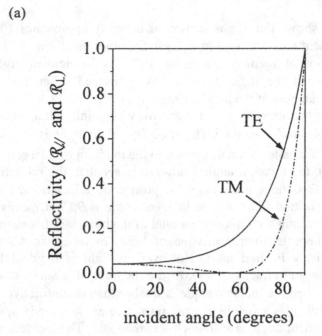

(b)

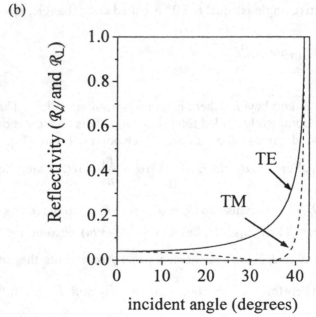

Figure 1-7. Intensity reflectivity vs. the incident angle. (a) Light is incident from the air (n=1) to the glass (n= 1.5).; (b) light is incident from the glass (n=1.5) to the air (n= 1).

incident from the medium of a larger refractive index to the medium of a smaller refractive index. It does not occur in the reverse direction. The three cases for $\theta_1 < \theta_c$, $\theta_1 = \theta_c$, and $\theta_1 > \theta_c$, are shown in Fig. 1-8.

(a)

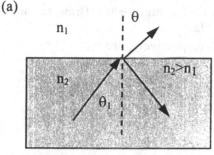

Refraction;$(\theta_1 < \theta_c, \theta_2 < 90^0)$

(b)

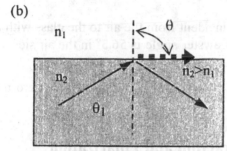

No refraction; $(\theta_1 = \theta_c, \theta_2 = 90^0)$

(c)

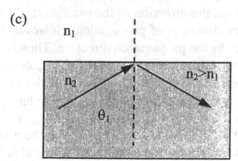

No refraction; $(\theta_1 > \theta_c)$

Figure 1-8. Illustration of reflection and refraction for (a)$\theta_1 < \theta_c$, (b) $\theta_1 = \theta_c$, (c) $\theta_1 > \theta_c$.

5.1.7 Brewster Angle

Figs. 1-7 (a) and (b) also show that at a certain angle, the reflection is zero for TM polarization. This angle is called Brewster angle, which can be calculated from Eq. (1-38a). For zero reflection

$$n_2 \cos\theta_i - n_1 \cos\theta_t = 0 \qquad\qquad (1\text{-}44)$$

Combined with the law of refraction, $n_1 \sin\theta_i = n_2 \sin\theta_t$, we

obtain $\tan\theta_i = \dfrac{n_2}{n_1}$ and $\tan\theta_t = \dfrac{n_1}{n_2}$. Therefore the Brewster angles, θ_{Bi} and

θ_{Bt}, in medium 1 and medium 2, respectively, are

$$\theta_{Bi} = \tan^{-1}(\frac{n_2}{n_1}) \qquad\qquad (1\text{-}45a)$$

$$\theta_{Bt} = \tan^{-1}(\frac{n_1}{n_2}) \qquad\qquad (1\text{-}45b)$$

For example, when light is incident from the air to the glass with n=1.5 or from the glass to the air, the Brewster angle is 56.3° in the air side and 33.7° in the glass side.

For TE polarization, Eq. (1-39a) gives no such angle for zero reflection, so there is no Brewster angle.

5.2   Reflection, Refraction and Polarization

For an electromagnetic wave, the direction of the oscillating electric field is usually perpendicular to the direction of propagation. However, there are many directions perpendicular to the propagation direction. Those directions are actually on a plane. If the direction of electric field randomly varies along on such plane, then we call this wave unpolarized. If the direction of electric field is fixed and along the x-axis, we call it linearly polarized and x-polarized, assuming that the propagation direction is along z-axis. The linearly polarized wave could be y-polarized for a wave with the direction of electric field along the y-axis. The polarization direction could also be along directions other than the x-axis or y-axis. In addition, it is possible that the direction of electric field regularly rotates on the plane perpendicular to the propagation direction. If the strength of the rotating field does not change, then this wave is circularly polarized. If the rotating field varies regularly, then it is called elliptically polarized. Fig. 1-9 schematically shows the above different polarization situations.

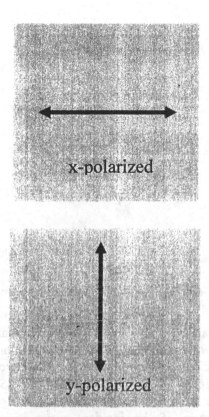

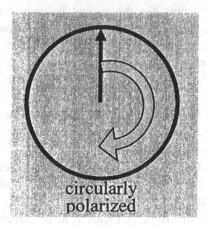

Figure 1-9. Different polarization situations. (continued)

Figure 1-9. Different polarization situations.

Unpolarized and circularly polarized waves can be represented by an x-polarized wave and a y-polarized wave: $E_x\hat{x} + E_y\hat{y}$ or any two other perpendicularly polarized waves. Fig. 1-6 shows that the electric field is decomposed into two components: TM polarization and TE polarization. If the incident wave is TM polarized only, then the reflection and transmission are completely given by Eqs. (1-38a) and (1-38b). Similarly, for the TE-polarized wave, the reflection and transmission are given by Eqs. (1-39a) and (1-39b). However, if the incident wave is unpolarized, then the reflection and transmission are the mixed behaviors of TE-polarized and TM-polarized waves. We can decompose the unpolarized wave into TE and TM waves, calculate the reflection and transmission of the TE and TM waves separately, then combine the reflected TE wave and the TM wave by adding their amplitude vectors. Combine the transmitted TE and TM waves in a similar fashion.

For a circularly polarized and unpolarized wave, the decomposed TE and TM components should be equal. However, when the incident wave is not normal to the boundary, the reflections of TE and TM waves are not the same and as a result, the TE and TM components of the reflected wave are not equal. Similarly, the transmitted wave has different amounts of TE and TM components. In this case they are called partially polarized wave. An extreme case is when the light is incident at the Brewster angle. The TM wave is then completely transmitted. Thus the reflected wave is only TE-polarized, as shown in Fig. 1-10. However, the transmitted wave contains both the TE and TM waves. If the incident wave has equal amount of TE and TM components, the transmitted wave should have more TM component than TE component.

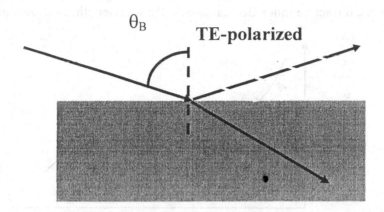

Figure 1-10. The reflected wave is TE-polarized only when the incident angle is equal to the Brewster angle (θ_B).

5.3 Dispersion

It is well known that a prism will divide the white light into different colors, as shown in Fig. 1-11. The reason for this phenomenon is because the refractive index of a material varies with the wavelength of the light. As a consequence, the refractive angle depends on the wavelength, as described by the following equation.

$$\frac{d\theta}{d\lambda} = \frac{d\theta}{dn}\frac{dn}{d\lambda} \qquad (1-46)$$

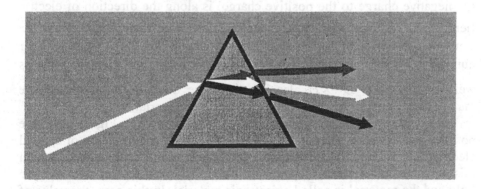

Figure 1-11. The beam of white light incident from the left side of the prism is divided into several beams of different colors.

The variation of the refractive index with wavelength is called dispersion. Usually, the refractive index decreases with the wavelength, as shown in Fig. 1-12.

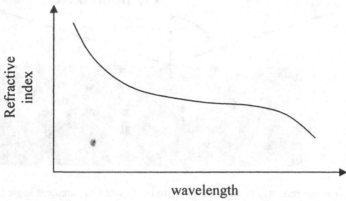

wavelength

Figure 1-12. A general variation of refractive index with wavelength.

Therefore, the short-wavelength component of light has a larger refractive angle than the long-wavelength component of light. As shown in Fig. 1-11, the blue-color light is bent more than the red-color light.

5.4   Isotropy and Anisotropy

Equation (1-18), $\bar{P} = \varepsilon_0 \chi_e \bar{E}$, states that an external field will cause polarization, which describes the separation of positive and negative charges. For crystals of good symmetry, as illustrated in Fig. 1-13 (a), the separation of positive and negative charges is independent of the direction of applied electric field. In addition, the dipole direction (the arrow line pointing from the negative charge to the positive charge) is along the direction of electric field. Then the polarization $\bar{P}$ has the same direction as the electric field $\bar{E}$. Thus the proportional constant, electric susceptibility χ_e, is only a scalar number and so the dielectric constant, $\varepsilon = \varepsilon_0(1 + \chi_e)$ is a scalar number as well. In this case, $\bar{D} = \varepsilon \bar{E}$ also has the same direction as $\bar{E}$ field. The material with such property is called an isotropic material.

For crystals with asymmetric structures, as illustrated in Fig. 1-13(b), the positive and negative charges will not separate along the direction of applied electric field due to different bonding forces among the x- or y-directions. In this situation, polarization depends on the direction of the incident electric field and the material is called anisotropic material. In this case, the value of χ_e depends on the direction of applied electric field. The dielectric constant, $\varepsilon = \varepsilon_0(1 + \chi_e)$, also depends on the direction of applied electric field. As a result, the polarization $\bar{P}$, the electric field $\bar{E}$, and the field $\bar{D}$ are not necessarily parallel. In addition, because the value of ε now depends on the direction of electric field, the refractive index $(= \sqrt{\mu\varepsilon} / \sqrt{\mu_0\varepsilon_0}$) is not a constant. The anisotropic property will result in the phenomenon of birefringence to be described in the following section.

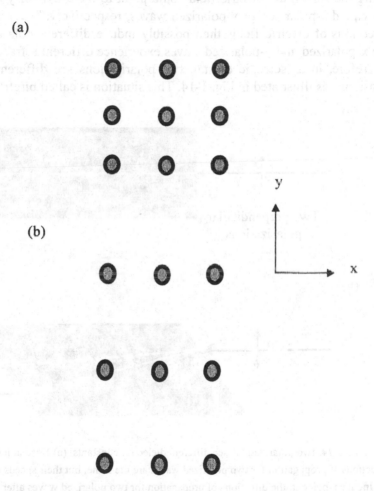

Figure 1-13. (a) isotropic: when electric field exists, charge separation is the same for x and y directions due to symmetric structure along both directions. (b) anisotropic: Because atom spacing along y-direction is larger, resulting in less bonding force for this direction, charges are easier to move along y-direction. Thus when electric field exists, charge separation along y-direction is larger than along x-direction.

5.5   Birefringence

For anisotropic materials, the refractive index depends on the direction of electric field. A propagating electromagnetic wave has two independent and perpendicular polarizations. For example, if the EM wave is propagating along the z-axis, the electric field could point to the x-axis or y-axis. These are called x-polarized or y-polarized waves, respectively. The two different directions of electric fields then possibly induce different polarizations, so the x-polarized and y-polarized waves experience different refractive indices. Therefore, in anisotropic crystal, two polarizations see different dielectric constants, as illustrated in Fig. 1-14. This situation is called birefringence.

(a)

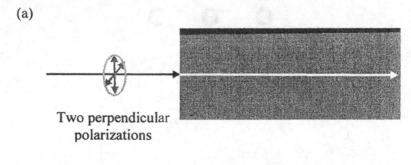

Two perpendicular
polarizations

(b)

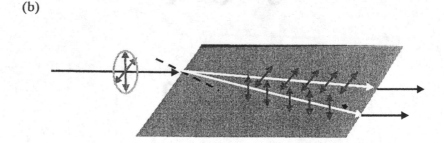

Figure 1-14. two polarizations see different dielectric constants. (a) Normal incident: the directions of propagation for two polarized waves are the same, but their speeds are different. (b) Inclined incident: the directions of propagation for two polarized waves after transmission are different. Also, their speeds are different.

5.6   Nonlinear Phenomenon

According to Eq. (1-18), $\vec{P} = \varepsilon_0 \chi_e \vec{E}$, the polarization is proportional to the electric field. However, nature is not that simple. Just as a spring's extension is sometimes not linearly proportional to the external force, the field-induced separation of positive and negative charges is also sometimes not linearly proportional to the field strength. As a result, the polarization $\bar{P}$ is related to the electric field $\bar{E}$ in a more complicated way.

$$\bar{P} = \varepsilon_0[\chi_e\bar{E} + \chi_2\bar{E}\bar{E} + \chi_3\bar{E}\bar{E}\bar{E} + \cdots] \tag{1-47}$$

The second term, third, and subsequent terms in the right hand side of the above equation are nonlinear terms that cause nonlinear phenomena.

Several nonlinear phenomena such as self-phase modulation, soliton effect, parametric process, Raman scattering, Brillouin scattering, to name a few, are often observed in optical fibers. Those phenomena can result in new frequency components in addition to the input signal. Some frequencies are very close to the input signal frequency and degrade the signal purity, but some could be useful for signal processing or transmission.

6 OPTICAL FIBER AND LIGHT GUIDING

In modern optical-communication systems, optical fibers are used as the media for light propagation. An optical fiber has a typical cross section shown in Fig. 1-15. The core in the center and the cladding surrounding the core are the two most important parts that determine the optical properties of the fiber. They are made from glass of different refractive indices. The core has the index n_1 which is larger than the index of cladding n_2 for light guiding purpose. Fiber with such an index variation is called *step-index fiber*, as illustrated in Fig. 1-16.

6.1 Light Guiding in Optical Fiber

According to the analysis in the previous section, when light is incident from the material of larger index to the material of smaller index, total reflection may occur. Therefore, if light is incident from the core to the cladding with an angle larger than the critical angle, total reflection occurs. As illustrated in Fig. 1-17, light traveling along the red arrow is incident on the boundary between the core and the cladding at the critical angle. In this case total reflection occurs and so light gets reflected and travels along the same direction as the red arrow. For light incident on the boundary between the core and the cladding with an angle larger than the critical angle, as shown by the purple arrow, total reflection occurs again. Thus light will not go into the cladding. Instead, it will propagate within the core region. On the other hand, if light is incident at the angle larger than the critical angle, as shown by the yellow arrow, part of the light will transmit to the cladding region and part of the light will be reflected back to the core region. However, the light reflected back to the core region will meet the boundary

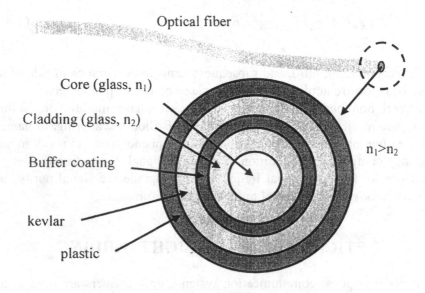

Figure 1-15. Cross section of optical fiber.

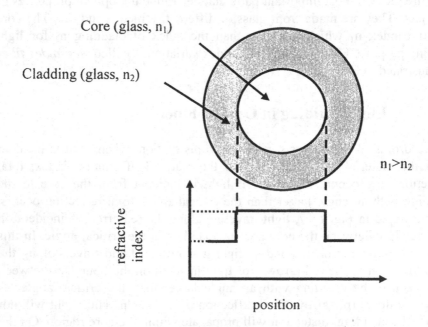

Figure 1-16. Index variation of a step-index fiber.

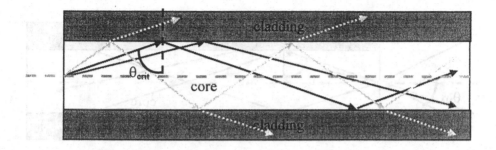

Figure1-17. Propagation of light rays in the optical fiber at different propagation directions.

again and transmission/reflection will occur once more. After many times, the light remaining in the core region becomes very weak. Therefore, for light propagating in the fiber over a long distance, only the light with the incident angle larger than the critical angle remains in the fiber.

Given the effect stated above, for light incident at the entry of the fiber, only those rays confined within a given cone angle are able to travel in the optical fiber for a long distance. (See Fig.1-18) After they enter the fiber, they will experience total reflection at the boundary of the core and the cladding. The maximum entry angle of those rays is defined as the *acceptance angle*. Rays with an entry angle larger than the acceptance angle do not experience total reflection. Thus, after they meet the boundary several times, the portion remaining in the core region becomes weak. This is shown by the yellow arrows in Figs. 1-17 and 1-18.

From the geometry shown in Fig.1-18, we know that the ray entering the fiber at the acceptance angle θ_{NA} will be incident on the boundary of the core and the cladding at the critical angle. Thus this angle should satisfy the following relation

$$\sin \theta_{NA} = (n_1^2 - n_2^2)^{1/2} \tag{1-48}$$

where n_1 and n_2 are the refractive indices of the core and cladding, respectively. This value is also defined as the *numerical aperture (NA)* of the fiber, $NA = \sin\theta_{NA}$, which is used to describe the capability of an optical fiber to collect light. The larger the numerical aperture, the more light rays the optical fiber can collect.

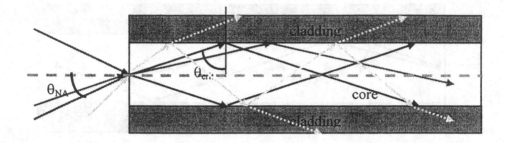

Figure1-18. Geometry of light rays entering the optical fiber (from left) and propagation path of light rays in the optical fiber.

6.2   Single-Mode and Multi-Mode Fibers

The analysis presented in the previous section is based on ray optics, where light is assumed to consist of many rays traveling in different directions. A better approach to use in analyzing the properties of an optical fiber is to consider light acting as wave. In wave-based analysis, Maxwell's equations must be solved.

Matching the core-cladding boundary conditions for electric and magnetic fields, guiding modes can be obtained. Due to the complicated mathematical procedures, the derivation is not shown here.

In general, some fibers have only one guiding mode. These are called single-mode fibers. Multi-mode fibers have multiple guiding modes. The number of guiding modes depends on the refractive-index profile, the core radius, and the wavelength. For step-index fibers, the parameter $V = \dfrac{2\pi a}{\lambda}\sqrt{n_1^2 - n_2^2}$ plays a very important role in determining the guiding modes. In this equation a is the radius of the core; n_1 and n_2 are refractive indices of the core and the cladding of the fiber respectively. For V < 2.35, there is only one guiding mode. Therefore, the single-mode fiber usually has small core radius, while the multi-mode fiber has a large core radius. Fig. 1-19 shows typical profiles of single-mode and multi-mode fibers. The fibers usually have a diameter of 125±2 μm, including both the core and the cladding. The diameter of the core of the single-mode fiber and the multi-mode fiber is about 9 μm and 50 μm, respectively.

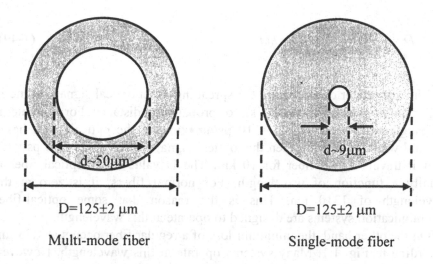

D=125±2 μm   D=125±2 μm

Multi-mode fiber   Single-mode fiber

Figure 1-19. Typical profiles of single-mode fiber and multi-mode fiber.

6.3 Modal Dispersion, Material Dispersion, and Waveguide Dispersion

For multi-mode fibers, many modes of light can be guided. Each mode has its own group velocity that is different from others, so signals carried by each mode arrive at the destination at different times, meaning that a signal of short-duration in time spreads out temporally at the destination. This phenomenon is called *modal dispersion*, which reduces the operational data rate of signals. For example, assume that signals are in the form of optical pulses. Each pulse has the duration of 10 ps with pulses separated at 100 ps for an operational data rate of 10 Gb/s. In multi-mode fiber, each pulse consists of many modes. Thus modal dispersion causes the pulse to spread out temporally. At the destination, the pulse duration becomes much wider than 10 ps. If the pulse duration becomes more than 100 ps, then neighboring pulses will overlap one another and cannot be distinguished from one another. Therefore, 10 GHz operation is not possible. Due to modal dispersion, multi-mode fibers can transmit signals typically below 1 Gb/s only. For single-mode fiber, the potential data rate can be more than 40 Gb/s.

Although single-mode fiber has no modal dispersion, it still has material dispersion and waveguide dispersion, which will cause optical pulse to spread out, too. However, this type of spread-out is much less than that caused by modal dispersion. To describe material dispersion and waveguide dispersion, the dispersion parameter D is used.

$$D = -\frac{\lambda}{c}\frac{d^2 n}{d\lambda^2} \quad ps/nm \bullet km \tag{1-49}$$

This parameter characterizes the spreading of an optical signal in time in ps per nm of spectral width and km of propagation distance. For example, if the single-mode fiber has $D = 10$ ps/nm•km and the optical pulse has a spectral width of 0.1 nm, then this optical signal will be at least 10 ps wide after it travels in the fiber for 10 km. The D value of an optical fiber is usually a function of wavelength. For normal fibers, it is zero at the wavelength of 1310 nm. This is the reason that some optical-fiber communication systems are designed to operate at this wavelength.

On the other hand, the minimum loss of a regular fiber occurs at 1.55 μm according to Fig. 1.1. Many systems operate at this wavelength. However, the dispersion at 1.55 μm is not zero. Therefore, some systems use special fibers that have the zero-dispersion shifted to 1.55 μm for optimal operation of data transmission. The fibers with zero-dispersion at 1.55 μm are called *dispersion-shifted fibers*. The dispersion-shifted fibers have special designs of the core and the cladding so that the waveguide dispersion compensates for the material dispersion to achieve zero dispersion at 1.55 μm. In the systems using dispersion-shifted fibers, the signals are transmitted at the minimum loss and zero dispersion.

The variation of dispersion parameter D with wavelength is illustrated in Fig. 1-20, where both normal fibers and dispersion-shifted fibers are shown. The total dispersion is the addition of material dispersion and the waveguide dispersion. The material dispersion is the same for both the normal fibers and the dispersion-shifted fibers.

6.4   Loss or Attenuation of Optical Fiber

An ideal transmission system for signals will be lossless. Then the signal can be transmitted to any place without the loss of signal level. Unfortunately, such systems do not exist. Every transmission system has a certain amount of loss that causes the power level of the signal to attenuate. The optical-fiber communication system also has loss, but its loss is known to be the smallest because current technology makes the loss of optical fibers very small. The loss of a fiber is usually characterized by the attenuation constant α. If the power of light launched at the input of a fiber of length L is P(0) and the power of the light from the output of the same fiber is P(L), then P(0) and P(L) can be related through the attenuation constant α.

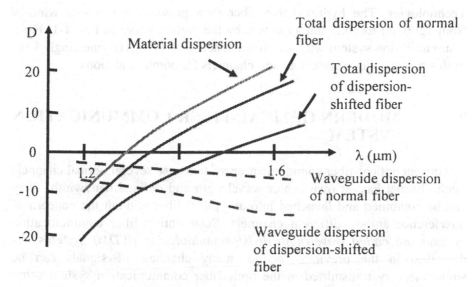

Figure 1-20. Variation of dispersion with wavelength.

$$P(L) = P(0) \exp(-\alpha L) \tag{1-50}$$

The attenuation constant is commonly referred to as the fiber loss and is usually expressed in the units of dB/km.

$$\alpha = -\frac{10}{L} \log(\frac{P(L)}{P(0)}) \text{ (dB)} \tag{1-51}$$

For a uniform optical fiber, the attenuation constant is independent of the fiber length. On the other hand, the attenuation constant is a function of wavelength. For light in the ultra-violet (UV) range, Rayleigh scattering and UV absorption are strong. These effects decrease with wavelength. In the infrared (IR) range, the IR absorption above a wavelength of 2 μm is also strong. The IR absorption decreases with decreasing wavelength. Thus the variation of fiber loss with wavelength behaves like the curve shown in Fig. 1-1.

The top curve shown in Fig. 1-1 has an absorption peak at 1385 nm. This peak is due to the absorption of the OH radical. The top curve shows that there are two low-loss windows. One is at 1310 nm with the loss level of 0.4 dB/km. The other is at 1550 nm with the loss level of 0.2 dB/km. If the OH radical is eliminated from the fiber material, the absorption peak at 1385 nm will disappear. Such a hydroxyl-free fiber has been manufactured by Lucent

Technologies. The hydroxyl-free fiber then provides a low-loss window from 1300 nm to 1625 nm as shown by the bottom curve in Fig. 1-1. For a communication system with a channel spacing of 100 GHz, one single fiber of this type provides more than 450 channels for communications.

7   MODERN OPTICAL-FIBER COMMUNICATION SYSTEM

Current optical-fiber communication systems use several optical channels. Each channel has its own center wavelength and defined bandwidth. They can be combined and launched into an optical fiber without the concern of interference among different channels. Such optical-fiber communication systems are *called wavelength division multiplexing (WDM)* systems. As described in the previous section, many channels of signals can be simultaneously transmitted in the optic-fiber communication system using one single fiber. The method used to send those channels into one fiber thus requires certain types of system layout. In addition, a variety of components are required in the system to incorporate those signals into the fiber system.

Fig. 1-21 shows a possible configuration of the optic-fiber communication system with three optical channels. XMT refers to the transmitter that converts the electronic signal to the optical signal. Before XMT, the signal is electronic. As shown in the plot, many electronic signals are combined through an electronic multiplexer. After the electronic multiplexer, the signals are at a very high bit rate and cannot be easily transmitted through conventional coaxial cables or electrical wires. Therefore, they have to be converted to optical signals in order to be transmitted in an optical fiber. Current systems are designed such that each optical channel carries 10 Gbs or 40 Gbs of optical signals. However, many more optical signals can be simultaneously transmitted in an optical fiber, so an optical multiplexer is used to combine several optical channels to be launched into one optical fiber. The system shown in Fig. 1-21 uses only three channels. However, as explained above, 450 channels can be simultaneously transmitted if necessary.

Although optical fiber has very small loss, an optical signal is still attenuated over a very long distance, requiring the signal to be amplified in order to reach the destination. The symbol "AMP" in the middle of the plot in Fig. 1-21 stands for the optical amplifier that is used to amplify optical signals. Fig. 1-21 shows only one optical amplifier, but more may be added between the launching source and the destination if required.

When the optical signal arrives at the other end of the optical fiber, they have to be directed toward each specified destination. Optical signals in different optical channels therefore have to be spatially separated to different directions using an optical demultiplexer. The optical signal in each channel is then converted back to an electronic signal using a receiver, designated by "RCV" in Fig. 1-21. The signals are then further divided to low-frequency electronic signals using an electronic demultiplexer.

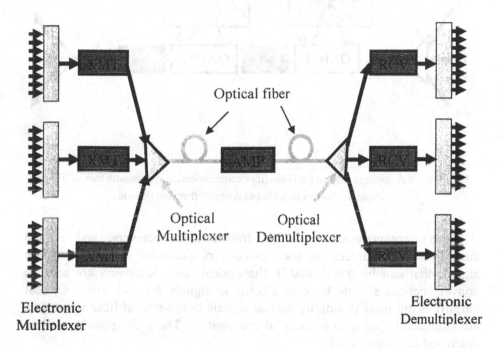

Optical fiber

Optical Multiplexer

Optical Demultiplexer

Electronic Multiplexer

Electronic Demultiplexer

Figure 1-21. A configuration of the optic-fiber communication system with three optical channels.

Figure 1-22 shows another possible configuration of the optic-fiber communication system with two additional channels that can be added or dropped from the system. In fact, each channel can be added or dropped from the system based on the customer's actual capacity requirement. Fig. 1-23 shows a ring network that has the same capability. Even more complicated systems than those shown above are possible.

In order to construct optical-fiber communication systems, many types of optical components are needed. The components used in the current optical-fiber communication systems can be categorized into two major parts: active components and passive components.

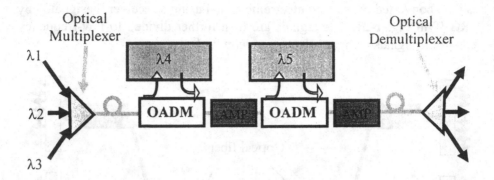

Figure 1-22. A configuration of the optic-fiber communication system with two additional channels that can be added or dropped from the system.

Active components consist of transmitters, receivers, and optical amplifiers. Transmitters are used to convert electronic signals to optical signals that can be transmitted in the optical fiber. Receivers are used to convert optical signals back to electronic signals for end users. Optical amplifiers are used to amplify optical signals in the optical fiber when they are attenuated after long distance of propagation. These components will be discussed in Chapters 2-5.

Passive components are those that require no external power. There are many such components, including:
– coupler, combiner/splitter;
– fixed and tunable filters;
– isolator;
– circulator;
– attenuator;
– wavelength division multiplexer/demultiplexer;
– optical add/drop multiplexer/demultiplexer;
– optical switch;
– optical crossconnect;
– wavelength router;
– wavelength converter.
These components will be described in Chapters 6-8.

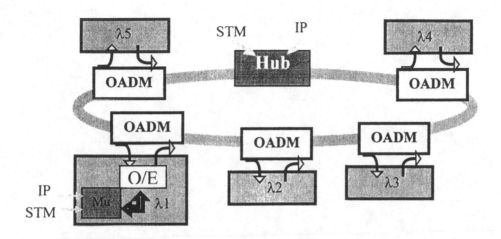

Figure 1-23. A ring configuration of the optic-fiber communication network. Each channel can be added or dropped from the system.

NOTES

1. The contents are described in a simple way. If readers are interested in detailed physics of electromagnetic waves, the particle nature of light, interaction of light with transparent materials, and so on, please see the references listed below.

REFERENCES

1. Shiji: 史記 (in Chinese).
2. Song's book: 宋・武經總要 (in Chinese).
3. Tang's book: 唐・通典 (in Chinese).
4. Beiser, Arthur, *Concepts of Modern Physics*. McGraw-Hill, Inc. 1995.
5. Born, Max and Wolf, Emil, *Principles of Optics*. Pergamon Press, 1959.
6. Yariv, Amnon and Yeh, Pochi, *Optical Waves in Crystals*. John Wiley & Sons, 1984.
7. Haus, Hermann A. *Waves and Fields in Optoelectronics*. by Prentice-Hall, Inc., 1984.
8. Yariv, Amnon, *Optical Electronics in Modern Communications*. 5/ed, Oxford University Press, 1997.
9. Agullo-Lopez, Fernando, Cabrera, Jose Manuel, and Agullo-Rueda, Fernando, *Electrooptics: Phenomena, Materials, and Applications*. Academic Press, 1994.

Chapter 2

TRANSMITTERS (I) --- LIGHT SOURCES

1.   INTRODUCTION

In an optical communication system, what is transmitted in the system is light that in turn carries signals. Therefore, the light source is a key component for optical communication systems. In this chapter, we will introduce the principle of light emission. Because lasers generate better quality of light than conventional light sources, we will then introduce the principle of lasers. Semiconductor lasers are particularly useful in modern fiber optic communication systems because of their compact size and efficient power conversion. Therefore, the operating principles of semiconductor lasers are discussed, followed by an introduction to several types of semiconductor light sources that are commonly used in modern fiber optic communication systems.

2.   PRINCIPLE OF LIGHT EMISSION

What is light? This question has puzzled human beings for thousands of years. A few meaningful points of view have been proposed. Light as a "Ray" is the first concept used to quantitatively describe the behavior of light. Ray propagation equations can precisely predict the trajectory of light propagation. In 17th century, light as a "particle" was used to predict the transfer of light from one place to another. However, the "particle" point of view was not able to explain the slower speed of light in water than in the air. Thus more physicists tended to use "wave" point of view to explain the behavior of light. Huygen's principle of wave and interference patterns from double slits by Thomas Young's experiment in 1801 provided a very good explanation and evidence for the wave behavior of light. However, in the 20th century, new discoveries like photoelectric effect made by Einstein and in 1923 the Compton effect by Compton, again proved that light also exhibits a "particle" behavior. Today we believe that light exhibits characteristics of both a "wave" and a "particle".

Generating light however is also a key issue. Before knowing the details of the nature of light, people already knew several ways to generate light, including burning charcoal, lumber, oil, or candles. After electricity was widely available, the electric light bulb became the common way to generate light. However, without knowing the physical principles involved, light emission was on a trial and error basis. Even the great inventor Edison failed more than 100 times before a good material for light bulb was discovered. Nowadays, we understand the physics of light emission much better, so many efficient ways of light emission have been invented.

2.1   Quantum Physics

The most important physics for light emission is quantum physics. It includes two parts. First, the energy of light is quantized. Second, the energy of an atomic system is quantized.

Because the energy of light is quantized, light carries energy in a multiple of a certain energy unit. This unit energy is equal to hv, where v is the frequency of light wave. As explained in Chapter 1, this is the particle nature of light. The light that carries the unit energy is called photon. When an atomic system interacts with light producing an exchange of energy, it must be in the way that emits or absorbs the light in multiples of photon energy units, hv.

The quantization of energy levels in an atomic system means that the energy levels are like the steps of a ladder. For example, the hydrogen atom has the energy levels given by the formula:

$$E_n = -\frac{13.6}{n^2} \text{ eV} \tag{2-1}$$

The energy levels are plotted in Fig. 2-1. The lowest energy level is called ground state. The other energy levels are called excited states. The electron of the hydrogen atom can stay on one of the energy levels. However, when the electron transits from the level n=1 to the level n=2, it has to absorb the energy ΔE.

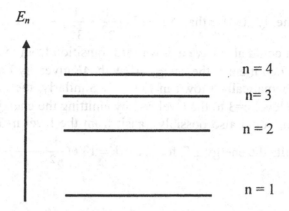

Figure 2-1. A schematic of energy levels of hydrogen atom.

$$\Delta E = 13.6(1 - \frac{1}{2^2}) = 10.2 \text{ eV}. \qquad (2-2)$$

Such transition is schematically shown in Fig. 2.2.

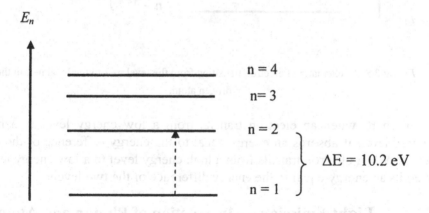

Figure 2-2. A schematic of electron transition from the level n=1 to the level n=2 in the hydrogen atom.

Similarly, when the electron transits from the level n=1 to the level n=3, it has to absorb the energy ΔE, while $\Delta E = 13.6(1 - \frac{1}{3^2}) = 12.09$ eV. The electron could also possibly transit from the level n=2 to the level n=3. Then it absorbs the energy ΔE, for that $\Delta E = 13.6(\frac{1}{2^2} - \frac{1}{3^2}) = 1.89$ eV.

The electron could also have a downward transition from the level n=2 to the level n=1. Then it emits the energy ΔE with ΔE given by Eq. (2-2). Such transition is schematically shown in Fig. 2.3. Similarly, the electron could transit from the level n=3 to the level n=1 by emitting the energy $\Delta E = 12.09$ eV. The electron could also possibly transit from the level n=3 to the level n=2. Then it emits the energy ΔE, for that $\Delta E = 13.6(\frac{1}{2^2} - \frac{1}{3^2}) = 1.89$ eV.

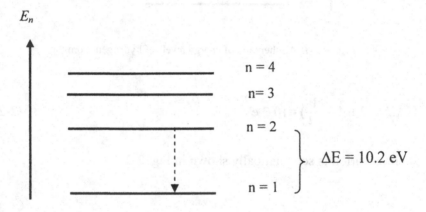

Figure 2-3. A schematic of electron transition from the level n=2 to the level n=1 in the hydrogen atom.

In short, when an electron transits from a low energy level to a high energy level, it absorbs an energy equal to the energy difference of the two levels. When electron transits from a high energy level to a low energy level, it emits an energy equal to the energy difference of the two levels.

2.2   Light Emission --- Interaction of Photon and Atomic Systems with Energy Levels

So how does the electron absorb or emit the energy for the transition? When an electron transits from a low energy level to a high energy level, where does the required energy for upward transition come from? When an electron transits from a high energy level to a low energy level, where does the energy that electron emits go? One possible way is that an electron absorbs the energy of a photon or gives the energy to a photon. As shown schematically in Fig. 2-4, when a photon with the energy equal to the energy difference of two energy levels of an atomic system encounters an electron, it can be absorbed by the system and cause the electron to transit from a low energy level to a high energy level. Similarly, when an electron transits from a high energy level to a low energy level, a photon is emitted. This photon has the energy equal to the energy difference of the two corresponding energy levels.

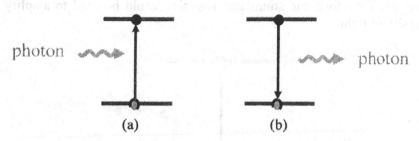

Figure 2-4. (a) a photon is absorbed by the atomic system, causing the electron to transit from a low energy level to a high energy level; (b) the electron in the atomic system transits from a high energy level to a low energy level, emitting a photon.

For most of materials, their atomic systems are more complicated than a hydrogen atom and their energy levels cannot be simply described by Eq. (2-1). Nonetheless, the transition of electrons from one energy level to another behaves in the same way. Therefore, from the viewpoint of quantum physics, light emission is due to the transition of electrons from high energy levels to low energy levels in an atomic system.

3.   PRINCIPLE OF LASERS

With this background in quantum physics, we are now ready to discuss the principle of lasers.

3.1   Spontaneous Emission and Stimulated Emission

When an electron is in an excited state, it will eventually transit to the ground state or a lower energy state. There are two possible ways for the electron to transit from an excited state to a low energy state. Fig. 2-5(a) shows the first way where a photon is directly emitted due to the transition of an electron from a high energy level to a low energy level. This process is called spontaneous emission because it needs no other photons to induce the transition. A second way is called stimulated emission, as shown by Fig. 2-5(b). In this process, the transition of an electron from a high energy level to a low energy level is induced by an incoming photon. After this transition two photons are emitted. One is the original incoming photon, the other is generated from the downward transition of the electron.

The stimulated emission tells us that a photon can cause the electron to transit from a high energy state to a low energy state by emitting another photon. With stimulated emission, the additional photon emitted is in phase with the incoming photon. Also both photons will propagate toward the same direction. Therefore, the stimulated emission could be used to amplify the intensity of light.

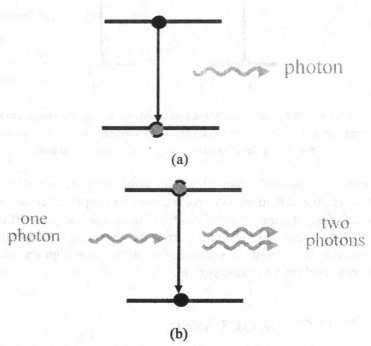

Figure 2-5. (a) spontaneous emission; (b) stimulated emission.

3.2   Absorption vs. Stimulated Emission

On the other hand, according to the description in the previous section, we also know that a photon can be absorbed by the atomic system and causes the electron to transit from a low energy state to a high energy state. So that begs the following question, when a photon is incident on the atomic system, will it be absorbed and cause electron to transit upward, or will it cause the stimulated emission so that electron transits downward for the emission of an additional photon? The condition is illustrated in Fig. 2-6.

The answer is simple. When the electron is at the low energy level, then the photon is absorbed. If the electron is at the high energy level, the incident photon will cause stimulated emission. However, the material usually has many such atomic systems, just like a glass of water having many H_2O molecules. Some may have electrons at the high energy levels and some have electrons at the low energy levels. Then when the light is incident on the material, whether the photon will be absorbed or cause stimulated emission is not clear.

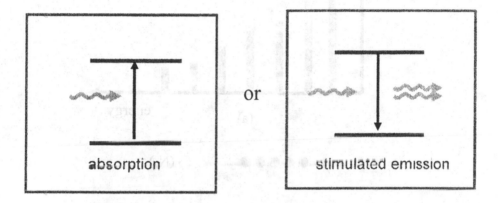

Figure 2-6. When a photon is incident on the atomic system, will it be absorbed or cause stimulated emission?

To answer this question correctly, we have to know the ratio of atomic systems with electrons at high energy levels to those with electrons at low energy levels. If there are more electrons at high energy levels than at low energy levels, then stimulated emission is more likely to happen and vice versa for absorption. However, according to thermodynamics, the number of electrons in high energy states must be smaller than in low energy states. The situation is illustrated in Fig. 2-7.

As shown in Fig. Fig. 2-7 (b), the number of electrons in a high energy level is smaller than those in a low energy level. In thermal equilibrium, the ratio of the population is usually given by the following equation.

$$\frac{N_2}{N_1} = e^{-(E_2 - E_1)/kT} \tag{2-3}$$

Therefore, there are more electrons at low energy levels than at high energy levels in thermal equilibrium. As a result, more absorption than stimulated emission will happen. The net effect of the material is the absorption of light.

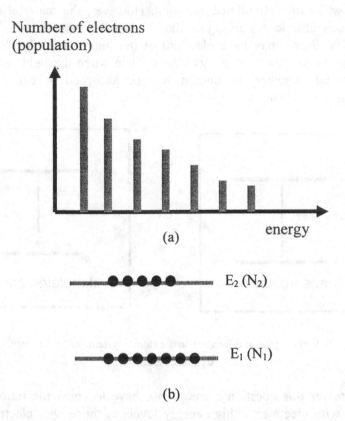

Figure 2-7. (a) Illustration of the distribution number of electrons vs. energy. (b) Number of electrons (N_2) in high energy level (E_2) is smaller than that (N_1) in low energy level (E_1).

3.3   Population Inversion

In order to achieve a net effect of stimulated emission, the number of electrons at high energy levels must be larger than those at low energy levels. That is $N_2 > N_1$. Such a condition is called population inversion because the population is the inverse of thermal equilibrium.

For a laser to work, three conditions are required. First, the laser gain material must have quantization levels in population inversion so that stimulated emission will happen. Stimulated emission is the necessary process for light amplification. In fact, the acronym laser stands for "Light Amplification by Stimulated Emission of Radiation". Second, there must be an excitation system to keep the gain material in population inversion because population inversion does not happen in thermal equilibrium. Third, a laser cavity is required to make light circulate therein so that light could obtain sufficient gain to overcome loss. The laser cavity is usually formed using mirrors. Fig.2-8 shows the schematic of a laser system.

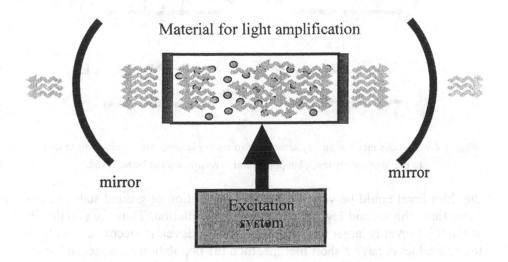

Figure 2-8. Schematic of a laser system. It consists of three major parts: (1) gain material for light amplification; (2) cavity formed by mirrors; (3) excitation system to maintain the gain material in population inversion.

Population inversion is the key point for light amplification. However, it is not easy to achieve population inversion. For a system with two energy levels shown in Fig. 2-9 (a), the high energy level has less electrons than the low energy level. When photons with energy equal to the energy difference of the two levels is incident on the system, they will be absorbed and so some electrons in low energy level transit to the high energy level, as shown in Fig. 2-9(b). With more photons incident on the system, more electrons transit to the high energy level. However, when electrons in the high energy levels are large in number, the photon could also cause electrons to transit to the low energy level. With very intense light incident on the system, the population in the two levels could be very close, as shown in Fig. 2-9(d). However, it is not possible to exceed a ration of 1 and achieve population inversion by light incident on the two-level system.

To achieved population inversion, one needs at least three energy levels. Fig. 2-10 shows a three-level system and a four-level system that can potentially produce population inversion. For a three-level system, the excitation light has energy equal to the energy difference of the ground state (first level) and the third level. With very strong excitation, the population at

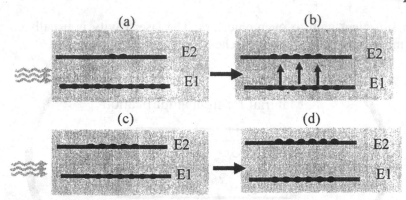

Figure 2-9. Photons incident on a system with two energy levels, causing electrons to transit to the high energy level, but population inversion cannot be achieved.

the third level could be very close to the population of ground state. At the same time, the second level is still at a low population. Thus the population at the third level is larger than that at the second level. If electrons staying at the second level have a short lifetime, then the population at the second level could remain low at all times. Population inversion is hence easily achieved. For a four-level system, the situation is similar. The excitation now causes electrons to transit to the forth level. Electrons at the forth level then quickly transit downward to the third level. At strong excitation, numerous electrons are excited to the forth level and quickly move to the third level. Because the second level has relatively low population, population inversion is achieved between the second and the third level. In this case, the electrons at the second level also have a short lifetime in order to maintain the population inversion.

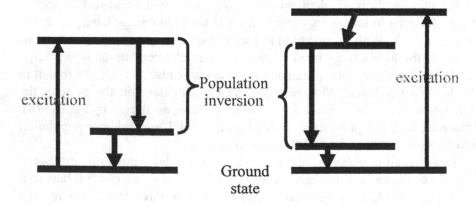

Figure 2-10. A schematic showing that population inversion can be achieved at three-level and four-level systems.

3.4 Laser Rate Equation with Only Spontaneous Emission

The behavior of laser action can be quantitatively analyzed in the following. Referring to Fig. 2-11, as electrons are excited to the energy level E2, they will transit downward to the energy levels E1 and E0 due to spontaneous emission. Therefore, the population N_2, defined as the number of electrons at the energy level E2 per unit volume, will vary with time according to the following equation

$$-\frac{dN_2}{dt} = \frac{N_2}{\tau_{sp}(2 \to 1)} + \frac{N_2}{\tau_{sp}(2 \to 0)} \tag{2-4}$$

where $\tau_{sp}(2 \to 1)$ and $\tau_{sp}(2 \to 0)$ are the spontaneous lifetimes associated with E2→E1 transition and E2→E0 transition, respectively. The inverse of the spontaneous lifetime is referred to as spontaneous transition rate. $1/\tau_{sp}(2 \to 1)$ = A_{21} and $1/\tau_{sp}(2 \to 0) = A_{20}$.

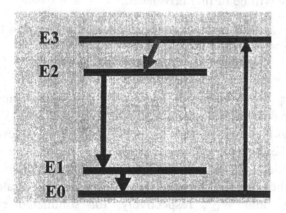

Figure 2-11. A four-level system used to describe population variation.

Eq. (2-4) can be further written using a single lifetime τ_{sp2}.

$$-\frac{dN_2}{dt} = \frac{N_2}{\tau_{sp2}} \tag{2-5}$$

where $1/\tau_{sp2} = [1/\tau_{sp}(2 \to 1)] + [1/\tau_{sp}(2 \to 0)]$.

From Eq. (2-5), the population N2 will be proportional to exp(-t/τ_{sp2}). Then the spontaneous emission of light will have a time variation of field as follows.

$$e(t) = E_o \cos \omega_o t \cdot e^{-t/\tau} \qquad (2-6)$$

This equation's Fourier transform gives the spectrum of light. Therefore, the spectral variation of the light is given by

$$\text{Intensity(function of frequency)} \propto |E(\omega)|^2 \propto \frac{1}{(\omega - \omega_o)^2 + (\frac{1}{\tau})^2} \qquad (2-7)$$

Therefore, the spectrum has a Lorentzian line shape with the full width at half maximum (FWHM) $\Delta v = \frac{1}{\pi \tau}$. This line-width is associated with the lifetime. If there are other mechanisms that influence the lifetime τ_{sp2}, then the line-width Δv will be further broadened.

$$\Delta v = \frac{1}{\pi} \frac{1}{\tau_2} = \frac{1}{\pi} (\frac{1}{\tau_{sp2}} + \frac{1}{\tau_c} + \frac{1}{\tau_p} + ...) \qquad (2-8)$$

where τ_c and τ_p are lifetimes because of other mechanisms that cause electrons to have downward transition.

When light with energy equal to the energy difference of level E2 and E1 is incident on the system, then light-induced transition also occurs. Such transition includes stimulated emission and absorption. They are proportional to the intensity of light. The induced transition rate per atom is defined as W_{21}' and W_{12}', respectively, for stimulated emission and absorption. W_{21}' and W_{12}' can be represented as

$$W_{21}' = B_{21}\rho(v) \text{ (for transition from E2 level to E1 level)} \qquad (2-9a)$$

$$W_{12}' = B_{12}\rho(v) \text{ (for transition from E1 level to E2 level)} \qquad (2-9b)$$

where $\rho(v)$ is the energy density of light per unit frequency and is related to light intensity (I) by

$$I = \frac{c\rho(\nu)}{n} \tag{2-10}$$

c is the speed of light and n is the refractive index.

Taking into account both the stimulated emission and spontaneous emission, we have the total downward transition rate equal to $N_2(B_{21}\rho(\nu) + A_{21})$. For upward transition, it is only induced by light, so the total upward transition equal to $N_1B_{12}\rho(\nu)$. Here again, N2 is the number of electrons at the energy level E2 per unit volume and N1 is the number of electrons at the energy level E1 per unit volume. At equilibrium, the total downward transition must equal the total upward transition. That is

$$N_2(B_{21}\rho(\nu) + A_{21}) = N_1B_{12}\rho(\nu) \tag{2-11}$$

From Eq.(2-11), the energy density of light can be derived and is given by

$$\rho(\nu) = \frac{A_{21}}{B_{12}e^{h\nu/kT} - B_{21}} \tag{2-12}$$

This formula will be identical to the energy density of light from black body radiation by assigning $B_{12} = B_{21}$ and

$$\frac{A_{21}}{B_{21}} = \frac{8\pi n^3 h\nu^3}{c^3} \tag{2-13}$$

The results tell us that, with the same intensity of light, stimulated emission rate per atom is equal to the absorption rate per atom. Also, the stimulated emission rate is related to the spontaneous emission rate. Because $A_{21} = 1/\tau_{sp}$ (2→1), combining Eq. (2-9), Eq. (2-10), and Eq. (2-13), we have the following transition rate W_i induced by light.

$$W_i = \frac{\lambda^2 I}{8\pi n^2 h\nu\tau_{sp}} \tag{2-14}$$

Here the lifetime τ_{sp} (2→1) is simply written as τ_{sp}. Each transition is associated with a line shape to explain the possible variation of transition with frequency, Eq. (2-14) becomes

$$W_i(v) = \frac{\lambda^2 I}{8\pi n^2 h v \tau_{sp}} g(v)$$ (2-15)

$g(v)$ represents the line shape function.

If light is incident on a material with the transition rate given by Eq. (2-15), it will be either absorbed or amplified, depending on the population difference $N_2 - N_1$. The absorbed or amplified light power per unit volume is then given by

$$\frac{P}{V} = (N_2 - N_1) W_i h v$$ (2-16)

If the material is a very thin slab with a thickness of dz, as shown by Fig. 2-12, the absorbed or amplified power of light is given by

$$P = I(z+dz) \cdot A - I(z) \cdot A$$ (2-17)

The volume V is given by

$$V = A \cdot dz$$ (2-18)

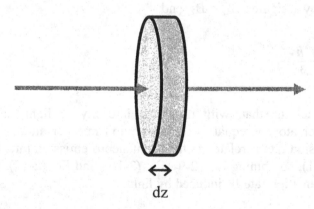

$\longleftrightarrow$
dz

Figure 2-12. A thin slab of material with the light-induced transition rate given by Eq. (2-15)

Substituting Eq. (2-15), Eq.(2-17) and Eq.(2-18) into Eq. (2-16), we obtain

$$\frac{dI}{dz} = (N2 - N1)\frac{c^2}{8\pi n^2 v^2 \tau_{sp}} g(v)I \tag{2-19}$$

Eq.(2-19) is more often written as

$$\frac{dI}{dz} = \gamma I \tag{2-20}$$

With the coefficient γ given by

$$\gamma = (N2 - N1)\frac{c^2}{8\pi n^2 v^2 \tau_{sp}} g(v) \tag{2-21}$$

Eq.(2-20) can be easily solved with the solution

$$I_{out} = I_{in} e^{\gamma z} \tag{2-22}$$

Eq. (2-22) says that if the light propagates in a thick material, its intensity will exponentially decay or grow, depending on the coefficient given by Eq.(2-21). If $N_2-N_1 < 0$, then γ is negative. The material is absorptive. The intensity of light will decay. The coefficient γ is usually written as $-\alpha$. α is called absorption coefficient with the unit of cm^{-1}. Eq. (2-22) becomes

$$I_{out} = I_{in} e^{-\alpha z} \tag{2-23}$$

If $N_2-N_1 > 0$, then γ is positive. This is the case of material with population inversion. Then light is amplified. Its intensity exponentially grows. The coefficient γ is then usually written as g. g is called gain coefficient with the unit of cm^{-1}. Eq. (2-22) becomes

$$I_{out} = I_{in} e^{gz} \tag{2-24}$$

Materials in thermal equilibrium have $N_2 - N_1 < 0$, so they are absorptive. Some pumping mechanisms are necessary to change the material from the thermal equilibrium to a state of population inversion in order to have gain for light amplification.

3.5   Laser Rate Equation with Stimulated Emission

To maintain population inversion for stimulated emission, there must be a pumping mechanism. Referring to Fig. 2-13, with the inclusion of pumping and stimulated emission, the rate equation for population N_2 becomes

$$\frac{dN_2}{dt} = R_2 - \frac{N_2}{\tau_2} - (N_2 - N_1)W_i(v) \tag{2-25}$$

where R_2 is the pumping rate, meaning the number of electrons optically induced to energy level E2 from ground state per second; τ_2 is the lifetime of electron staying at the energy level E2. This lifetime includes all of the mechanisms that cause electrons to have downward transition.

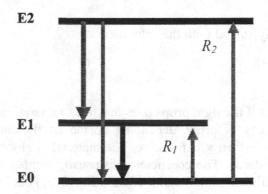

Figure 2-13. A three-level system used to explain laser rate equation.

The rate equation for population N_1 is given by

$$\frac{dN_1}{dt} = R_1 - \frac{N_1}{\tau_1} + \frac{N_2}{\tau_{sp}} + (N_2 - N_1)W_i(v) \tag{2-26}$$

where R_1 is the pumping rate, meaning the number of electrons optically induced to energy level E1 from ground state per second; τ_1 is the lifetime of electron staying at the energy level E1. This lifetime also includes all of the mechanisms that cause electrons to have downward transitions. τ_{sp} is the spontaneous-emission lifetime associated with the E2→E1 transition. The term $(N_2 - N_1)W_i(v)$ is the optically induced transition between the energy levels E2 and E1. If $N_2-N_1 > 0$, the transition is downward, so the population N_2 decreases and the population N_1 increases. $W_i(v)$ is given by Eq.(2-15).

At the steady state, the time variation vanishes. Then Eq. (2-25) and Eq.

(2-26) can be solved simultaneously to give the steady-state population N_2 and N_1 and yield the population difference $N_2 - N_1$.

$$N_2 - N_1 = \frac{\Delta N^0}{1 + I/I_s} \qquad (2\text{-}27)$$

where

$$\Delta N^0 = R_2 \tau_2 - (R_1 + \frac{\tau_2}{\tau_{sp}} R_2) \tau_1 \qquad (2\text{-}28)$$

$$I_s = \frac{8\pi n^2 h\nu \tau_{sp}}{[\tau_2 + (1 - \frac{\tau_2}{\tau_{sp}})\tau_1]g(\nu)} \qquad (2\text{-}29)$$

Eq. (2-27) shows that the population difference $(N_2 - N_1)$ decreases with the light intensity. According to Eq. (2-21), the coefficient γ is proportional to the population difference $(N_2 - N_1)$, so γ also decreases with the light intensity.

$$\gamma = \frac{\gamma_0}{1 + I/I_s} \qquad (2\text{-}30)$$

γ_0 is the γ coefficient in Eq. (2-21) with the population difference $(N_2 - N_1)$ replaced by ΔN^0. For $(N_2 - N_1) > 0$, γ represents the gain coefficient. For $(N_2 - N_1) < 0$, it becomes the absorption coefficient. The decrease of gain or absorption with the light intensity is called gain saturation or absorption saturation. The saturation behavior is shown in Fig. 2-14.

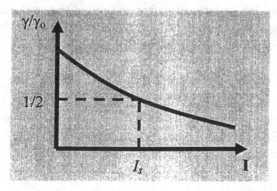

Figure 2-14. The saturation behavior of the γ coefficient.

γ_0 is the unsaturated gain or unsaturated absorption coefficient. As the intensity is very small, compared to I_s, the gain or absorption coefficient is very close to the unsaturated one. When the light intensity is at I_s, the gain or absorption coefficient decreases to only half of the unsaturated value. Further increase of the light intensity eventually leads this term to approach zero.

When $(N_2 - N_1) > 0$, the material has population inversion, so it has gain. Then, according to Eq. (2-27), ΔN^0 must be greater than zero. From Eq. (2-28), we know that, in order to have $\Delta N^0 > 0$, τ_2 must be larger than τ_1, and R_l should be as small as possible.

3.6   Laser

With the gain material in a cavity formed by two mirrors, light can bounce forward and backward with amplification. Fig. 2-15 illustrates a simple cavity configuration with two mirrors and its cavity filled with gain material. Note that the gain material also has internal loss due to impurity absorption or scattering of light.

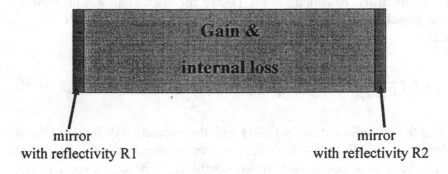

            mirror   mirror
    with reflectivity R1   with reflectivity R2

Figure 2-15. A simple cavity configuration formed by two mirrors with gain material within it.

As the light retains its intensity after one-round trip in the cavity, the gain will be exactly equal to the loss, which includes the internal loss and mirror loss. This is expressed by the following equation.

$$R1R2e^{2(\gamma-\alpha)l} = 1 \qquad\qquad (2\text{-}31)$$

where α is the internal loss; $R1$ and $R2$ are the mirror reflectivity; l is the cavity length. Eq. (2-31) can be further derived to give the threshold condition of the coefficient γ.

$$\gamma_t = \alpha - \frac{1}{2l}\ln(R1R2) \tag{2-32}$$

Therefore, the gain has to overcome the internal loss and the mirror loss.

When the laser is in the steady state, the light intensity is maintained at the value that causes the gain coefficient to saturate. The gain saturation then makes the gain coefficient drop to the value given by Eq.(2-32), as illustrated in Fig. 2-16.

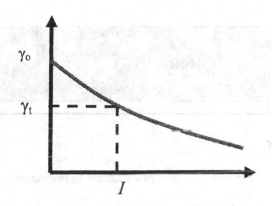

Figure 2-16. Gain saturation as a function of intensity. As the gain coefficient drops to the value of γ_t, the corresponding intensity is the light intensity in the cavity.

3.7   Laser Modes

For a laser cavity formed by two mirrors, as shown in Fig.2-17, the light wave oscillating between the two mirrors should satisfy the following condition.

$$nL = m\ (\lambda/2) \tag{2-33}$$

where n is the refractive index of the material in the cavity; L is the cavity length; λ is the wavelength of light and m is an integer. This condition assures that light wave has constructive interference for each round trip. Otherwise, the light wave in different round trips will cancel one another, leading to no light in the cavity. Therefore, Eq. (2-33) selects those wavelengths of light waves that can exist in the cavity. The selected wavelengths are

$$\lambda_m = 2nL/m \tag{2-34}$$

The selected light waves with the above wavelength are called longitudinal modes. The neighboring modes have the spectral separation given in terms of wavelength

$$\Delta\lambda = \lambda_{m+1} - \lambda_m = \lambda^2/(2nL) \tag{2-35}$$

or in terms of frequency

$$\Delta f = c/(2nL) \tag{2-36}$$

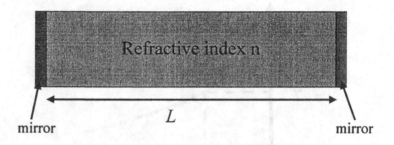

Refractive index n

mirror L mirror

Figure 2-17. A laser cavity formed by two mirrors.

From Eq. (2-35), we realize that the spectral separation is a function of cavity length. For typical lasers, the cavity length is much larger than the wavelength, so the spectral separation is very small. Thus there might be many longitudinal modes in the cavity. On the other hand, because there is always loss for each mode, those longitudinal modes will eventually vanish unless there is gain in the cavity. Therefore, only those longitudinal modes that situate within the line-shape of the gain can possibly exist in the cavity, as schematically shown in Fig. 2-18. However, due to the mode competition, not all of those modes will survive. Then the number of actual lasing modes is much less than the number of possible lasing modes discussed above. For the case of extreme competition, there might be only lasing mode.

In addition to the selection of longitudinal modes, the wave equation also requires the beam profile (transverse to the propagation direction) to satisfy certain distribution. Those selected beam profiles are called transverse modes. Usually the cavity is designed in the way that only the fundamental transverse mode exists. The intensity of the beam profile of the fundamental transverse mode usually has a Gaussian distribution, so the laser beam is often called a Gaussian beam. Its intensity distribution is given as follows.

$$I(r) = \frac{2P}{\pi d^2} e^{-2r^2/d^2} \tag{2-37}$$

where P is the power of the beam; d represents the spot size; r is the distance from the beam center.

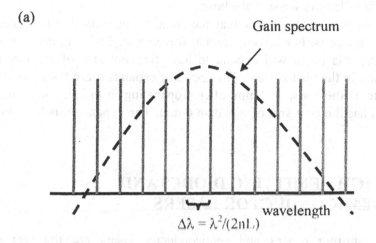

Figure 2-18. (a) Longitudinal modes that are within the gain line-shape. (b) The possible lasing spectrum that contains those lasing modes.

The spot size expands slightly with the propagation distance. Assuming that the beam propagates along the z direction and the minimum spot size is at the location of $z = 0$, the expansion of the spot size follows the equation

$$\omega(z) = \omega_0 \sqrt{1 + (\frac{z}{z_R})^2} \tag{2-38}$$

where ω_0 is the minimum spot size at z=0 and $Z_R = \dfrac{\pi\omega_o^2}{\lambda}$. The minimum spot size is called the waist of the beam.

From Eq. (2-38), we know that the smaller the waist is, the faster the beam divergence is. For a laser beam at wavelength 1.55 μm, if the waist is only 2 μm, this beam will expand to have the spot size of 2.5 mm after propagating in the air for 1 cm. The spot size expands more than 3 orders of magnitude. If the waist is 5 mm, after propagating in the air for 20 m, this beam still has the spot size of less than 6 mm. The expansion is less than 20 %.

4.   LIGHT EMITTING DIODES AND SEMICONDUCTOR LASERS

Light emitting diodes and semiconductor lasers are the two most common light sources used for optical communication systems because of their compact size and long operation lifetime. Both light emitting diodes and semiconductor lasers are made of semiconductors. In this section we will introduce their working principles and properties, beginning with the fundamental property of a semiconductor.

4.1   Property of Semiconductor

Semiconductors are materials that have better conductivity than insulators, but worse than conductors. The reason is because semiconductors have a small portion of electrons that can freely move among atoms. Although those freely moving electrons in semiconductors are fewer than those in conductors, their number is still quite large. In contrast, insulators have almost no freely moving electrons.

The difference among conductors, semiconductors, and insulators can be distinguished from the viewpoint of bandgap energy. As shown in Fig. 2-19, insulators have large bandgap energy, while semiconductors have their energy bandgap larger than zero, but approximately less than 3 eV. Conductors have no bandgap. The physical meaning of bandgap will be explained in the following.

For crystal materials, the atoms are naturally arranged in a rigid order, so the potential in the crystal is periodic. Putting the periodic potential into the Schrödinger equation, we can solve the energy levels in the atomic system of crystals. Those energy levels can be divided to two groups. The group of low energy levels is called valence band. The group of high energy levels is called conduction band. The two groups of energy levels are separated by

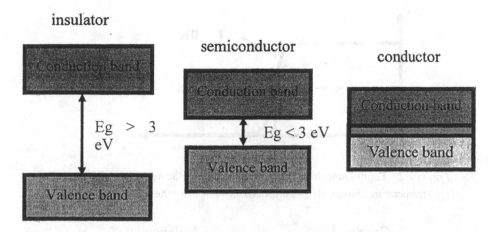

Figure 2-19. The difference among conductor, semiconductor, and insulator distinguished from the viewpoint of bandgap

bandgap energy, Eg, for insulators and semiconductors. The energy levels in the valence band are usually occupied by electrons, which are bonded to certain atoms. On the other hand, electrons in the conduction band can freely move among atoms in the crystal.

According to Fermi-Dirac statistics, each energy level of energy E is occupied by the electron with the probability

$$F(E) = \frac{1}{1 + e^{(E-E_f)/kT}}$$ (2-39)

The distribution function is plotted in Fig. 2-20. When the energy level has the energy E equal to Fermi energy E_f, the probability is 1/2. If the energy of the energy level is far below E_f, the probability is 1. On the other hand, if the energy is far above E_f, the probability is 0. The Fermi-Dirac distribution is also a function of temperature. As temperature approaches 0 K, it becomes a step function. That is, the probability is 1 for energy below E_f and 0 for energy above E_f.

For insulators, the conduction band and the valence band are widely separated. Then the energy levels in the valence band are far below the Fermi energy, E_f, which is approximately at the middle of bandgap. Also, the energy levels in the conduction band are far above the Fermi energy, E_f. As a result, almost every energy level in the valence band is occupied by an electron and the energy levels in the conduction band have negligible electrons. The situation is illustrated in Fig. 2-21. Therefore, insulators have almost no conduction carriers.

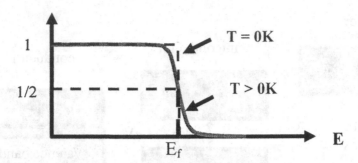

Figure 2-20. Fermi-Dirac distribution. Red solid line is the distribution for a certain
temperature above 0 K. The brown dashed line is for the temperature 0 K.

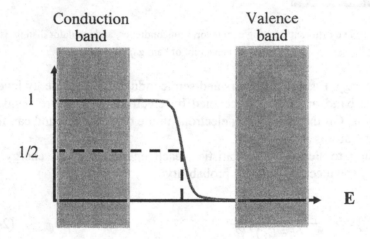

Figure 2-21. The relative position of conduction band, valence band and Fermi-Dirac
distribution in insulators.

For semiconductors, the conduction band and the valence band are close.
In this case some energy levels in the valence band are near the Fermi energy,
E_f. In addition, some energy levels in the conduction band are also close to
the Fermi energy, E_f. Therefore, some energy levels in the valence band are
not occupied by electrons and some energy levels in the conduction band are
occupied by electrons. That is, some electrons in the valence band have
obtained sufficient thermal energy and go to the high energy levels in the
conduction band. The situation is illustrated in Fig. 2-22. Therefore,
semiconductors have some conduction carriers.

For conductors, there is no bandgap. Therefore, there are lots of electrons
in the conduction band to give conductivity.

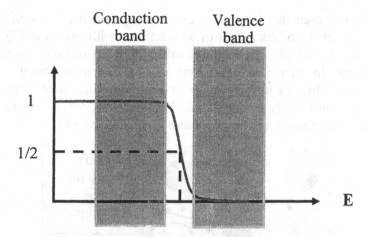

Figure2-22. The relative position of conduction band, valence band and Fermi-Dirac distribution in semiconductors.

4.1.1 Doping Effect on Semiconductors

A good property of semiconductors is that their conductivity can be changed by doping another type of atoms into the semiconductor. Fig. 2-23 illustrates the bonding situation of pure Si. Each Si atom is bonded to another four Si atoms because Si has four electrons in its outer most orbit. The pure Si without doping is called intrinsic. If one Si atom is replaced with an atom that has 5 outer electrons, then this impurity atom is also bonded to four Si atoms. However, there is one more electron left unbonded, as illustrated in Fig. 2-24. This electron can thus move freely among atoms. The Si crystal doped with impurity atoms such that there are more freely moving electrons than intrinsic Si is called n-type Si.

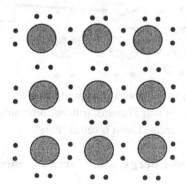

Figure. 2-23. Each Si atom is bonded to four other Si atoms, so each atom has eight electrons in the outer most orbit.

On the other hand, if one of Si atom is replaced with an atom that has 3 outer electrons, then this impurity atom is also bonded to four Si atoms. However, one of the orbits is lack of an electron, as illustrated in Fig. 2-25. The place that is lack of one electron is called "hole". This hole can be filled by an electron in its neighboring orbit, causing the neighboring orbit to become empty. This is equivalent to the movement of the "hole". Therefore, this hole can move freely among atoms. The Si crystal doped with impurity atoms so that there are freely moving holes is called p-type Si.

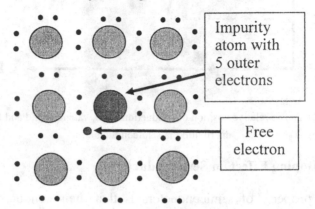

Figure 2-24. One of Si atom is replaced with an impurity atom that has 5 outer electrons. This impurity atom is also bonded to four Si atoms, with one free electron unbonded.

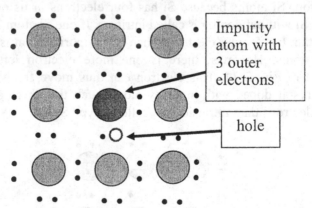

Figure 2-25. One of Si atom is replaced with an impurity atom that has 3 outer electrons. This impurity atom is also bonded to four Si atoms, with one orbit unfilled with electron. This unfilled orbit is called "hole".

4.1.2   Doping, Fermi-Energy, and Carrier Concentration

n-type Si has more free electrons than the intrinsic Si. This means that the Fermi level is closer to the conduction band, so the energy levels in the conduction band have more probability of being filled with electrons. For p-type Si, there are more holes than the intrinsic Si. Because "hole" means unfilled energy levels, the Fermi level is closer to the valence band. Fig. 2-26 shows the location of Fermi level relative to the conduction band and valence band for intrinsic, n-type, and p-type Si.

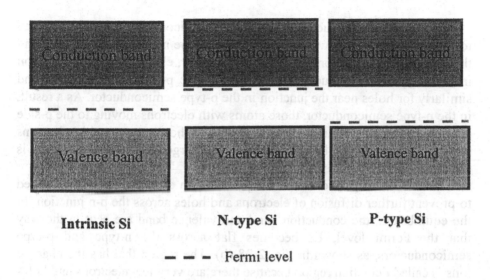

Intrinsic Si   **N-type Si**   **P-type Si**

— — — Fermi level

Figure 2-26. The location of Fermi level relative to the conduction band and valence band for intrinsic, n-type , and p-type Si.

In intrinsic Si, the number of electrons and the number of holes are equal. In contrast, the n-type Si has more electrons than holes, while p-type Si has more holes than electrons. The concentration of electrons (n) and the concentration of holes (p) follows the relation

$$n \cdot p = n_i^2 \tag{2-40}$$

where n_i is a constant. Electrons have negative charge and holes have positive charge. Both electrons and holes are carriers that can conduct current. The doped semiconductor, either n-type or p-type, is more conductive than the intrinsic one. For example, if $n_i = 10^{11}/\text{cm}^3$, the semiconductor is doped to have electron concentration $n = 10^{16}/\text{cm}^3$. Then according to Eq.(2-40), the hole concentration is $p = 10^6/\text{cm}^3$. This doped semiconductor has total carrier concentration $n + p \sim 10^{16}/\text{cm}^3$. For intrinsic semiconductor, $n = p = n_i = 10^{11}/\text{cm}^3$. The intrinsic semiconductor then has total carrier concentration $n + p = 2 \times 10^{11}/\text{cm}^3$. The carrier concentration of the doped semiconductor is almost five orders of magnitude larger than the intrinsic semiconductor. Because the conductivity is proportional to the carrier concentration, the doped semiconductor is more conductive than the intrinsic semiconductor.

4.2   p-n Junction

The n-type semiconductor and the p-type semiconductor can be connected to form a diode. As shown in Fig. 2-27, when the n-type semiconductor and the p-type semiconductor are connected together, electrons near the junction in the n-type semiconductor will diffuse to the p-type semiconductor and similarly for holes near the junction in the p-type semiconductor. As a result, in the n-type semiconductor, those atoms with electrons moving to the p-side become positively charged ions. Similarly, those atoms with electrons moving to the n-side become negatively charged ions. The situation is illustrated in Fig. 2-28 (a).

Due to the charged ions near the junction, an electric field is established to prevent further diffusion of electrons and holes across the p-n junction. In the equilibrium, the conduction band and valence band are bent in the way that the Fermi level, E_f, becomes flat across the n-type and p-type semiconductors, as shown in Fig. 2-28 (b). The region that has the charged ions is called depletion region because there are very few electrons and holes in this region.

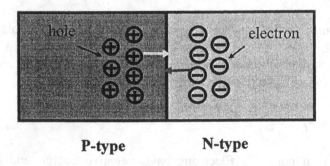

P-type   **N-type**

Figure 2-27. Electrons and holes diffuse across the p-n junction to the other side of semiconductor.

When the p-n junction is under reverse bias, as shown in Fig. 2-29, electrons and holes are pulled further away from the junction region. This will increase the depletion region and enhance the electric field in the depletion region, making it very difficult for electrons and holes to flow through. On the other hand, when the p-n junction is under forward bias, as

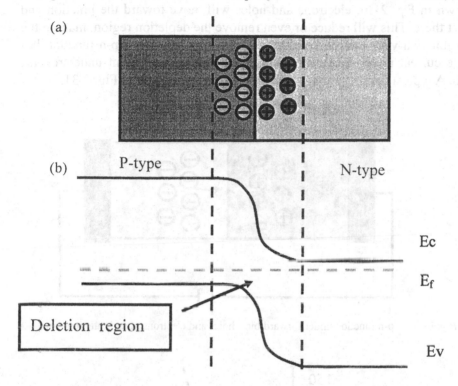

Figure 2-28. (a) After electrons and holes near the junction region diffuse to the other side, those atoms become charged ions. (b) Due to the charged ions near the junction, the conduction band and valence band are bent, leading to the Fermi level, E_f, flat across the n-type and p-type semiconductors.

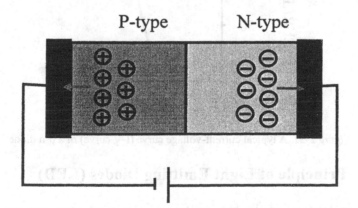

Figure 2-29. p-n junction under reverse bias: holes and electrons move away from the junction.

shown in Fig. 2-30, electrons and holes will move toward the junction and meet there. This will reduce or even remove the depletion region, making the p-n junction very easy to conduct current. That means the p-n junction has large current under forward bias and has very small current under reverse bias. A typical current-voltage curve (I-V curve) is shown in Fig. 2-31.

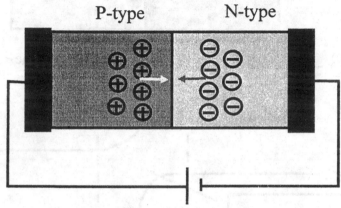

Figure 2-30. p-n junction under forward bias: holes and electrons meet in the junction.

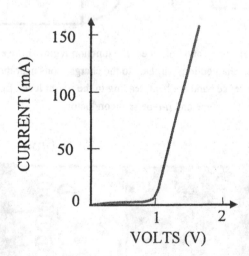

Figure 2-31. A typical current-voltage curve (I-V curve) of a p-n diode.

4.3   Principle of Light Emitting Diodes (LED)

After understanding the p-n junction of semiconductors, we are ready to study the working principle of light emitting diodes. The simplest structure of a light emitting diode contains only p-n junction. As explained before, when it is under forward bias, electrons and holes meet at the junction. They then recombine to emit photons. However, in order to increase light emission efficiency, another semiconductor with a lower bandgap energy is sandwiched between the n-type and p-type semiconductors. The structure is like the sketch shown in Fig. 2-32 (a). As this structure is under forward bias, the band diagram across the junction is shown in Fig.2-32 (b). Under forward bias, electrons are injected from the n-type semiconductor and holes are injected from the p-type semiconductor. Because the semiconductor sandwiched between the n-type and p-type semiconductors has smaller bandgap energy than the two sides, as shown in Fig. 2-32 (b), electrons and holes are confined in this region. Therefore, electrons and holes of high concentration simultaneously appear in this region and greatly increase the radiative recombination. The radiative recombination of electrons and holes means that electrons transit from the conduction band to the valence band and photons are emitted. The photon thus has the energy equal to the difference of the energy levels in the conduction band and the valence band. The structure with semiconductors of different bandgap energy is called hetero-structure. Nowadays, the center

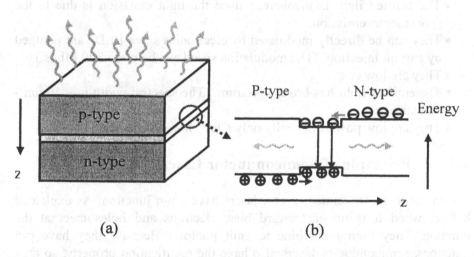

Figure 2-32. (a) A structure of LED with a semiconductor of low bandgap energy sandwiched between the n-type and p-type semiconductors. (b) The corresponding band diagram under forward bias.

semiconductor of low energy bandgap even has quantum wells to further increase the light emission efficiency.

In short, the working principle of the light emitting diode is as follows.
– It is under forward bias.

– Many electrons and holes meet simultaneously in the junction region.
– Electrons and holes recombine to emit photons.

4.3.1   Characteristics of LEDs

LEDs thus have the following characteristics:
• They are made from semiconductor materials.
• They are fabricated using semiconductor processing techniques.
• The wavelength of the emitted photon is approximately given by wavelength (λ, in μm) = 1.24/Eg(eV), where Eg is the bandgap of the semiconductor used to emit light.
• They are commonly made from the following Semiconductor materials

Material	Bandgap (eV)	Wavelength (μm)
GaAs	1.42	0.87
InP	1.35	0.92
InGaAsP	0.73 - 1.35	0.9 - 1.7

• The emitted light is incoherent since the light emission is due to the spontaneous emission.
• They can be directly modulated by electronics since LEDs are pumped by current injection. (The modulation speed can be up to 300 Mbps.)
• They are low cost.
• The emitted light has broad spectrum. (The spectral width is ~ 50 nm - 100 nm.)
• They are low power. (Usually only a few mW of light is generated.)

4.4   Principle of Semiconductor Lasers

Similar to LED, semiconductor lasers have a p-n junction. As explained before, when it is under forward bias, electrons and holes meet at the junction. They then recombine to emit photons. Because they have p-n junctions, semiconductor lasers also have the rectification property, so they are called laser diodes (LDs). In addition, like all other lasers, two more conditions are required: population inversion and a pair of parallel mirrors to form a linear cavity or several mirrors to form a ring cavity. Because population inversion is required, the population of electrons and holes in semiconductor lasers usually has to be larger than that in LEDs. This has led to the use of quantum-well structures in modern designs. The band diagram plotted in Fig. 2-33 is similar to that of Fig. 2-32 (b), but with quantum wells. . In this plot, there are three quantum wells. Under forward bias, electrons are injected into quantum wells from the N-cladding layer and holes are injected from the P-cladding layer. The thickness of quantum wells is usually less than 10 nm, so quantum wells have quantized energy levels as shown by the dashed line in Fig. 2-33. As a result, the transition is from the quantized level of quantum wells in the conduction band to the quantized level of quantum wells in the valence band. The number of quantum wells could range from one to five, depending on the design. With quantum wells, the gain coefficient of semiconductor lasers could easily exceed 100 cm^{-1}.

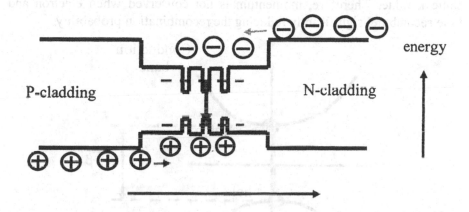

Figure 2-33. Band diagram of a semiconductor-laser material with quantum wells.

The quantum wells are usually not directly connected to the cladding layers. Between them, there is a region called separate confinement hetero-structure (SCH). The SCH layer typically has the thickness between 30 nm and 200 nm. The SCH layer provides two functions. First, carriers, including electrons and holes, have to traverse this region when they travel from the cladding layer to the quantum wells. This gives the injected carriers some time to relax their energy and settle down in the quantum wells. After their energy is reduced, they are better confined in the quantum wells. Second, the semiconductors with low bandgap energy usually have a high refractive index, so the entire region sandwiched between the N-cladding layer and the P-cladding layer forms a waveguide. Light is mostly confined in this region. Therefore, the necessary interaction of light and gain media for laser action is provided by such structure.

In practice, the cladding layer, SCH layer, and quantum wells are epitaxially grown on InP wafers or GaAs wafers. In order to guarantee the quality of grown crystal, two growth techniques are often used now. One is molecular-beam epitaxy (MBE). The other is metal-organic chemical vapor deposition (MOCVD) or called metal-organic vapor-phase epitaxy

(MOVPE).

Compared to other lasers, semiconductor lasers need one more condition satisfied. The semiconductor for light emission should have direct bandgap, which means that the semiconductor has the band structure schematically shown in Fig. 2-34. In the k-space, the conduction band minimum and the valence maximum have the same k-value. Therefore, the momentum ($= \hbar k$) of electron and hole are the same. When they recombine to emit photons, the momentum is conserved. In contrast, for indirect bandgap semiconductor, the conduction band minimum and the valence maximum do not have the same k-value. Therefore, momentum is not conserved when electron and hole recombine, significantly reducing the recombination probability.

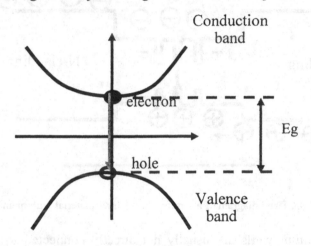

Figure 2-34. Band structure of direct-bandgap semiconductor.

Similar to LED, the emission wavelength is given by

$$\text{wavelength } (\lambda, \text{ in } \mu m) = 1.24/E \text{ (eV)} \qquad (2\text{-}41)$$

where E is the energy difference of the quantized level of quantum wells in the conduction band and the quantized level in the valence band.

4.4.1   Formation of Laser-Diode Cavity

The cavity of semiconductor lasers can be easily formed by the naturally cleaved facets. Because semiconductors are crystals, the cleaved facets along certain directions are particularly flat. The flatness is much better than artificially polished mirrors. In addition, because semiconductors usually have the refractive index as large as 3.5, the interface between the semiconductor and air thus gives the reflectivity of around 30 %. Because semiconductor lasers have very large gain coefficient, this reflectivity is sufficient to make light in the cavity oscillate. Similar to laser operation, the gain for semiconductor lasers has to overcome the losses.

$$\underbrace{e^{2GL}}\underbrace{e^{-2\alpha L}}\underbrace{R_1 R_2} = 1 \qquad\qquad (2\text{-}41)$$

gain   Internal   Mirror
loss   loss

This gives the threshold gain: $G_{th} = \alpha - \log(R_1 R_2)/2L$, where L is the cavity length. The gain coefficient in semiconductor lasers is approximately a linear function of injection current: $G = a(I - I_{tr})$. It thus gives the threshold current of semiconductor lasers. When the injection current is above the threshold, the output power of laser light increases significantly with injection current, as schematically shown in Fig. 2-35.

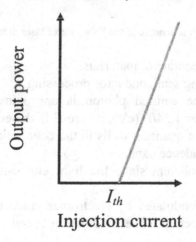

Figure 2-35. A typical L-I (light-current) curve of semiconductor laser.

Semiconductor lasers with naturally cleaved facets are usually called Fabry-Perot laser diodes because the two parallel cleaved facets form a Fabry-Perot cavity. The lasing modes are thus selected by the Fabry-Perot cavity. A schematic of the Fabry-Perot laser diode is shown in Fig. 2-36. The device length L is approximately between 300 μm and 1000 μm.

4.4.2 Characteristics of Fabry-Perot Laser Diodes

The characteristics of Fabry-Perot laser diodes are summarized as follows.

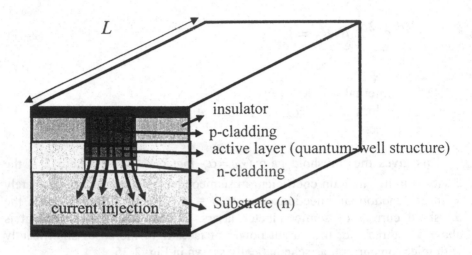

Figure 2-36. A schematic of the Fabry-Perot laser diode.

- They are made of semiconductor materials.
- They are fabricated using semiconductor processing.
- The wavelength of the emitted photon is approximately given by wavelength (λ, in μm) = 1.24/E (eV), where E is the energy difference of the quantized level of quantum wells in the conduction band and the quantized level in the valence band.
- The emitted light is coherent since the light emission is due to the stimulated emission.
- They can be directly modulated by electronics since laser diodes are pumped by current injection. (The modulation speed can be up to 10 Gbps.)
- There are usually multiple longitudinal modes. (Each mode has the line width of about 0.001 nm - 0.005 nm. The mode spacing is given by $\lambda^2/(2nL)$. The total spectral width of the multiple modes is about 1 nm – 8 nm.)
- The output power is in the range of a few mW to 100 mW.

5. SEMICONDUCTOR LASERS FOR OPTICAL COMMUNICATION

Because the Fabry-Perot laser diodes have multiple modes, they are not good for the applications in optical communication. Some modifications are therefore used to make laser diodes more suitable for optical communication. The following are some common types of semiconductor lasers used for this purpose.

5.1 Distributed Feedback (DFB) Lasers

Fig. 3-37 shows a cross section of a DFB laser along the cavity direction. A Bragg grating is formed along the cavity. The Bragg grating causes strong reflection at a certain wavelength. Therefore, DFB lasers oscillate at a single mode with line-width of ~ 50 kHz. Because of the narrow line-width, chirp is greatly reduced. In addition, due to single mode operation, no mode hopping occurs, so there is much less relative intensity noise (RIN).

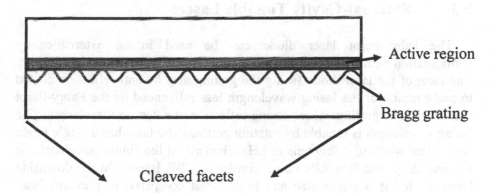

Figure 3-37. A schematic of the cross section of a DFB laser along the cavity direction.

5.2 Distributed Bragg Reflector (DBR) Lasers

Temperature and refractive indices change in DFB lasers when current passes through the active region and the Bragg grating.. This leads to the variation of wavelengths selected by the Bragg grating. To avoid the variation of lasing wavelength, another structure, called a distributed Bragg reflector (DBR) is used. A schematic of a DBR laser is shown in Fig.2-38. The Bragg grating does not extend over the entire cavity. Instead, the Bragg reflector now only occupies a portion of the cavity. In addition, the section with the Bragg reflector has no gain, so no current passes through the Bragg reflector. This means the temperature and refractive index remain constant. As a result, the lasing wavelength of DBR lasers has much less shift in comparison to the DFB lasers, while the other characteristics are similar to DFB lasers. The Bragg reflector is also used to cause strong reflection at a certain wavelength, so DBR lasers oscillate at a single mode with a line width of 0.0001nm. There is also less chirp and less RIN noise, compared to Fabry-Perot laser diodes.

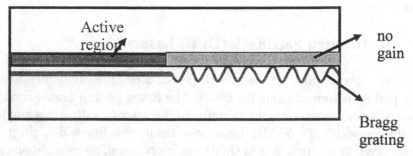

Figure 3-38. A schematic of the cross section of a DBR laser along the cavity direction.

5.3   External-Cavity Tunable Lasers

The Fabry-Perot laser diode can be used in an external-cavity configuration with a grating to become a tunable laser, as shown in Fig. 2-39. The facet of the laser diode facing the grating may be anti-reflection coated to make tuning of the lasing wavelength less influenced by the Fabry-Perot resonance. The grating causes strong reflection at a certain wavelength. The lasing wavelength is tunable by rotating grating. The laser has a single mode with a line width of a few tens of kHz. Because of the single wavelength, it has less chirp, and less RIN noise, similar to DBR lasers. On the downside however, it has a larger size and higher cost compared to previous laser diodes.

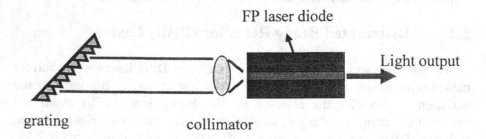

Figure 2-39. A schematic of an external-cavity tunable semiconductor laser.

5.4   Vertical Cavity Surface Emitting Lasers (VCSELs)

The laser beam of the above semiconductor lasers is emitted from the cleaved facet. They are sometimes also called edge-emitting lasers. The propagation direction is in parallel with the junction plane. The laser beam can only be measured when the laser diodes are cleaved apart. In addition, the laser beam usually has an elliptical profile because the emission aperture is not circular. The elliptic beam shape also makes the coupling of light into fibers difficult because the fiber profile is circularly symmetric.

To avoid the disadvantages of edge-emitting laser diodes, surface emitting lasers are developed. In particular, the vertical cavity surface emitting lasers (VCSELs) are commonly used in fiber optic communication. The side view and the top view of VCSELs are schematically shown in Fig. 2-40. The cross section of the layer structure shown in the side view is similar to that of the edge emitting laser diodes. The difference is that the p-cladding and n-cladding layers are now replaced by the mirror stacks, which are actually Bragg reflectors formed by many interlaces of materials with high-low refractive indices.

Again, the Bragg reflectors cause strong reflection at a certain wavelength. Thus the light can be reflected between the top p-mirror stack and the bottom n-mirror stack to form a vertical cavity. The QW active layer has gain, so light can oscillate. The light emits vertically from the cavity. The emitting aperture is usually circular, giving the light beam a circular beam profile.

VCSELs have the following characteristics that are mostly different from those of edge-emitting laser diodes.
- They have a single longitudinal mode, similar to DFB laser diodes because the lasing wavelength is selected by the Bragg reflector.
- VCSELs have low threshold currents, usually less than 5 mA. since the This is due to the Bragg reflector typically exhibiting a reflectivity near 100%. However, this also means the output power is small for the same reason, typically less than 1 mW. The low power limits their applications mainly in local area network (LAN).
- The light beam is circularly symmetric and has low divergence, making coupling to optical fiber is easy.
- The light has no particular polarization and so is good for fiber communication.
- Direct modulation by electronics is possible since it is also pumped by current injection.. The modulation bandwidth can be up to 2.4 GHz.
- The commonly available wavelengths are around 850 nm and 980 nm, good for short-range optical fiber communication. Development on 1.3 µm VSCELs is active, so commercial availability is expected within a

few years of this writing.

• There is no need to cleave apart the devices for beam measurement since the light is emitted vertically. Thus they are easily arrayed and tested.

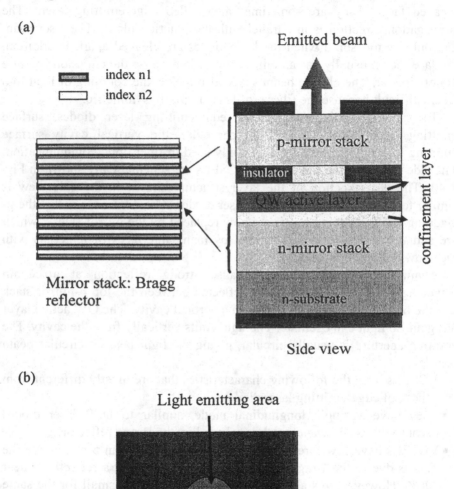

Figure 2-40. (a) Side view of vertical cavity surface emitting lasers. (b) Top view of vertical cavity surface emitting lasers.

NOTES

1. Readers who want to know more about quantum physics and Schrödinger equation are referred to Ref. 1.
2. Ref. 2 introduces the physics of lasers in a simple way. It is good to be an entry level. For detailed physics of lasers, please refer to Ref. 3.
3. For semiconductor physics and solid state physics, please refer to Refs. 4-6.
4. For the derivation of Fermi-Dirac statistics, please refer to Ref. 7.
5. Readers who want to know the physics and detail of semiconductor lasers in depths are referred to Refs. 8-9.

REFERENCES

1. Beiser, Arthur, *Concepts of Modern Physics*. McGraw-Hill, Inc. 1995.
2. Hecht, Jeff, *Understanding Lasers: an Entry-Level Guide*. IEEE Press, 1991.
3. Yariv, Amnon, *Optical Electronics in Modern Communications*. 5/ed, Oxford University Press, 1997.
4. Kittel, Charles, *Introduction to Solid State Physics*. John Wiley & Sons, Inc., 1976.
5. Ashcroft, Neil W., and Mermin, N. David, *Solid State Physics*. 1976.
6. Sze, S. M., *Semiconductor Devices: Physics and Technology*. John Wiley & Sons, Inc., 1985.
7. Reif, F., *Fundamentals of Statistical and Thermal Physics*. McGraw-Hill, 1965.
8. Agrawal, Govind P. and Dutta, Niloy K., *Semiconductor Lasers*. 2nd Ed., Van Nostrand Reinhold, 1993.
9. Yoshihisa Yamamoto, *Coherence, Amplification, and Quantum Effects in Semiconductor Lasers*. John Wiley & Sons, Inc., 1991.

Chapter 3

TRANSMITTER (II) --- MODULATORS

1.   INTRODUCTION

The signals transmitted in the optical communication system are carried by light. Therefore, the function of a transmitter is to convert the electronic signals to optical signals in which signals are carried by light. In Chapter 2, we learned how to generate light. Here we will describe how to put signals on the light. A transmitter module therefore includes two parts: a light source and modulator. Its block diagram is shown in Fig. 3-1.

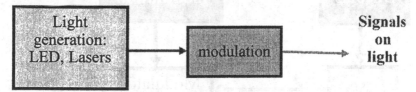

Figure 3-1. Block diagram of a transmitter module.

In practice, there are two ways to achieve the above purpose. They are direct modulation and external modulation. Direct modulation is illustrated in Fig. 3-2. Because laser diodes are pumped by current injection, they can be modulated directly through electronic signals. The AC signals (electrical modulation) and DC bias are combined through a bias T and then delivered simultaneously to the laser diode. The intensity of the light output then naturally varies according to the signals of electrical modulation. External modulation is illustrated in Fig. 3-3. The laser is only biased at a DC level, so the light output from the laser has no signal. Then the light beam is delivered to a modulator. This modulator has the absorption or loss varying with the signals of electrical modulation. Thus the light generation and modulation are accomplished at two different optical components. Both will be explained in this chapter.

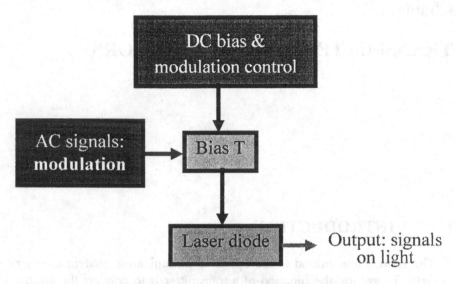

Figure 3-2. A transmitter module with direct modulation.

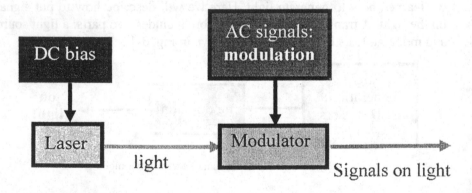

Figure 3-3. A transmitter module with external modulation.

2. DIRECT MODULATION OF LASER DIODES

When the laser diode is operated above the threshold current, the intensity of light is ideally a linear function of injection current. As shown in Fig. 3-4, when the DC-bias current is far above the threshold current (I_{th}), an AC modulation added to the DC bias will result in a proportional variation of light intensity. Although Fig. 3-4 shows that the modulation signal is sinusoidal, other types of modulation work similarly. The frequency response of the direct modulation can be analyzed in the following.

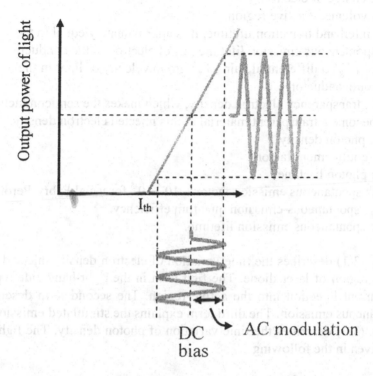

Figure 3-4. The response of light output due to the electrical modulation of injection current.

2.1   Rate Equation of Laser Diodes

Like other lasers, the electrons and light in laser diodes obey the rate equations. For easy formulation, the light is transformed to photon density. The rate equations are as follows.

$$\frac{dN}{dt} = \frac{I}{qV} - \frac{N}{\tau} - A(N - N_{tr})\Phi\Gamma \tag{3-1}$$

$$\frac{d\Phi}{dt} = A(N - N_{tr})\Phi\Gamma - \frac{\Phi}{\tau_p} + \beta_{sp}\eta_{sp}\frac{N}{\tau_e} \tag{3-2}$$

The definition of these symbols is listed below.

N : electron density.

I : injected current.

Q : charge of electron.

V : volume of active region.

τ : interband transition lifetime. It is approximately equal to the spontaneous emission lifetime (τ_e) of electron in the conduction band.

$A = a\,V_g$, a differential gain ; V_g : group velocity of light in the semiconductor.

N_{tr} : transparency electron density, which makes the semiconductor become a transparent material at this injected electron density.

Φ : photon density.

Γ : confinement factor.

τ_p : photon lifetime

β_{sp} : spontaneous emission factor ~ 10^{-4}-10^{-5} for usual Fabry-Perot LDs.

η_{sp} : spontaneous-emission quantum efficiency.

τ_e : spontaneous emission lifetime.

Eq. (3-1) describes the time variation of electron density injected into the active region of laser diode. The first term in the right-hand side represents the current injection into the active region. The second term describes the spontaneous emission. The third term explains the stimulated emission.

Eq. (3-2) describes the time variation of photon density. The light power P is given in the following

$$P = \Phi\,h\nu\,\sigma V_g \qquad\qquad\qquad\qquad (3\text{-}3)$$

where σ is the cross section of the active region. $h\nu$ is the energy of a single photon. The first term in the right hand side is the stimulated emission. The second terms explains the loss of light in the cavity. It includes the internal loss and the mirror loss. The third term is the contribution of spontaneous emission to the laser mode, which is very small and therefore often neglected.

If there is only DC bias, the electron density and photon density will first fluctuate and then enter steady state after some time (usually less than 1 μs). Before the steady state, the fluctuation is called the transient state, which is illustrated in Fig. 3-5.

After some time, the steady state will be reached. Then the time variation is zero, i.e., $dN/dt = 0$ and $d\Phi/dt = 0$. Eq. (3-1) and Eq. (3-2) become

$$\frac{I}{qV} - \frac{N}{\tau} - A(N - N_{tr})\Phi\Gamma = 0 \qquad\qquad (3\text{-}4)$$

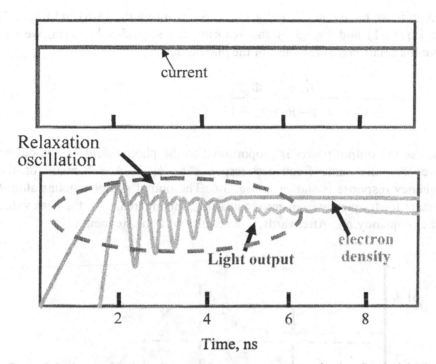

Figure 3-5. The transient behaviors of light output and electron density after the injection current is turned on.

$$A(N - N_{tr})\Phi\Gamma - \frac{\Phi}{\tau_p} + \beta_{sp}\eta_{sp}\frac{N}{\tau_e} = 0 \qquad (3\text{-}5)$$

From Eq. (3-4) and Eq. (3-5), we can obtain the steady-state solutions of electron density, N_o, and photon density, Φ_o, in terms of DC-bias current I_o.

2.2   Modulation of Laser Diodes[1]

When the injection current is modulated, we have $I = I_0 + ie^{j\omega t}$, where a sinusoidal modulation is assumed; i is the modulation amplitude and ω is the modulation frequency. With small-signal modulation, i.e., $i \ll I_o$, the electron density and the photon density will have similar formats.

$$N = N_o + ne^{j\omega t} \qquad (3\text{-}6)$$

$$\Phi = \Phi_o + \phi e^{j\omega t} \qquad (3\text{-}7)$$

Again, we have $n \ll N_o$ and $\phi \ll \Phi_o$. Substituting Eq. (3-6) and Eq. (3-7) into Eq. (3-1) and Eq. (3-2) and ignoring the second-order term, we can solve the small-signal response of the photon density.

$$\phi(\omega) = \frac{-(i/qV)A\Phi_o\Gamma}{\omega^2 - j\omega/\tau - j\omega A\Phi_o - A\Phi_o/\tau_p} \qquad (3\text{-}8)$$

Because the output power is proportional to the photon density, the output power has the same frequency response as $\phi(\omega)$. A schematic of the frequency response is shown in Fig. 3-6. The output power remains almost constant for frequency from zero to near ω_{os}, then increases to the peak value at the frequency ω_{os}. Afterwards, it decreases with the frequency.

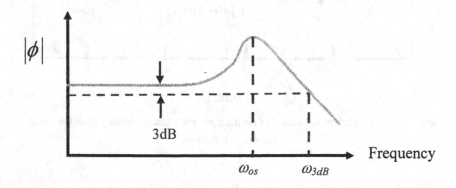

Figure 3-6. Frequency response of photon density (output power).

The characteristic frequency ω_{os} is a special property of the laser diode.

$$\omega_{os} = \sqrt{\frac{A\Phi_o}{\tau_p} - \frac{1}{2}(\frac{1}{\tau} + A\Phi_o)^2}$$

$$\sim \sqrt{\frac{A\Phi_o}{\tau_p}} = \sqrt{\frac{1 + A\tau_p\Gamma N_{tr}}{\tau\tau_p}(\frac{I_o}{I_{th}} - 1)} \qquad (3\text{-}9)$$

This frequency is called relaxation oscillation frequency. It is a natural resonance frequency of the laser diode. When the modulation frequency is near this frequency, the system resonance of the laser diode enhances the frequency response. Thus a peak response is observed in Fig. 3-6. The relaxation oscillation of a laser diode is usually somewhere between 1 GHz and 35 GHz, depending on the parameters of the laser diode. A larger relaxation frequency also results in a higher frequency response. Eq. (3-9) demonstrates that several factors influence the modulation response, including DC-bias level, spontaneous emission lifetime, photon lifetime (depending on cavity length), differential gain, a and so on. Fig. 3-7 illustrates the influence of the DC-bias level on the modulation response.

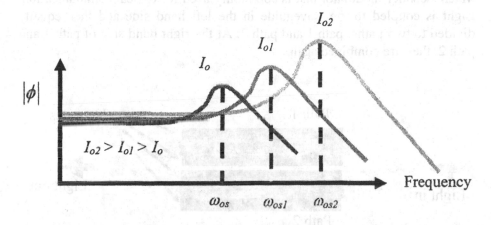

Figure 3-7. Modulation response for different DC-bias levels.

The analysis above assumes the small variation of injection current. For actual applications in optical communication, the modulation amplitude, i, may not be small. Then the frequency response will deviate from the response derived above. However, the small-signal analysis still provides a reasonably good approximation of the actual response.

2.3   Chirp [2,3]

In laser diodes, the refractive index varies with the carrier density. When modulation is applied to the laser diode, the injection current varies with time. This causes the carrier density to vary with time, leading to the time variation of refractive index. As a result, the length of the optical path and so the phase of the lasing mode varies with time. The time variation of phase then induces new frequency components and causes the frequency to vary with time. Chirp thus occurs and leads to the broadening of the laser line-width. For laser diodes, the line-width broadening induced by chirp is in the range of 100MHz to 1GHz/mA, which is approximately 0.01% - 0.001% of the center frequency.

3.   EXTERNAL MODULATION

3.1   Mach-Zehnder Modulator

For external modulation, the modulator and light source are not the same component, as shown in Fig. 3-3. One of the popular modulators is the Mach-Zehnder modulator. Fig. 3-8 shows a schematic of the waveguide-type Mach-Zehnder modulator that is commonly used for optical communication. Light is coupled to the waveguide in the left hand side and then equally divided to two paths: path 1 and path 2. At the right hand side of path 1 and path 2, they are combined again.

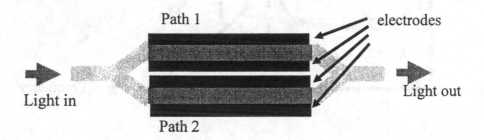

Figure 3-8. A schematic of waveguide-type Mach-Zehnder modulator.

In path 1 and path 2, the waveguide is surrounded by electrodes, so their refractive index can be modulated by the applied voltage. As a result of the index change, the phase of light passing through path 1 and path 2 can be changed. When lights passing through path 1 and path 2 are in phase at the right hand side, the light output is strong because of constructive interference. If they are out of phase, the light output is weak due to destructive interference. Therefore, the intensity of light at the output is modulated by applying voltages to the electrodes.

The key issue for the waveguide-type Mach-Zehnder modulator is how to change the refractive index in path 1 and path 2. In Chapter 1, we introduced anisotropic materials in which the refractive index depends on the direction of electrical field of the EM wave. The index of an anisotropic crystal is generally described by the index ellipsoid in the principal coordinate system:

$$\frac{x^2}{n_x^2} + \frac{y^2}{n_y^2} + \frac{z^2}{n_z^2} = 1 \tag{3-9}$$

where n_x is the refractive index that the EM wave experiences when the electrical field of the EM wave is along the x-direction. n_y and n_z are refractive indices for similar effects along the y- and z-direction, respectively.

For some crystals like $LiNbO_3$, the refractive index can be further changed by an applied electrical field. Then the equation of the index ellipsoid is changed to become the following form. [4]

$$(\frac{1}{n_x^2} + r_{1k}E_k)x^2 + (\frac{1}{n_y^2} + r_{2k}E_k)y^2 + (\frac{1}{n_z^2} + r_{3k}E_k)z^2$$
$$+ 2yzr_{4k}E_k + 2zxr_{5k}E_k + 2xyr_{6k}E_k = 1 \tag{3-10}$$

where $r_{ik}E_k$ denotes the following summation

$$r_{ik}E_k = r_{i1}E_x + r_{i2}E_y + r_{i3}E_z, \quad i = 1, 2, \ldots, 6 \tag{3-11}$$

r_{ik} is the electro-optic coefficients. It is usually given in the form

$$[r_{ij}] = \begin{pmatrix} r_{11} & r_{12} & r_{13} \\ r_{21} & r_{22} & r_{23} \\ r_{31} & r_{32} & r_{33} \\ r_{41} & r_{42} & r_{43} \\ r_{51} & r_{52} & r_{53} \\ r_{61} & r_{62} & r_{63} \end{pmatrix} \tag{3-12}$$

For example, the electro-optic coefficients for $LiNbO_3$ have the properties $r_{12} = -r_{22}$, $r_{23} = r_{13}$, $r_{42} = r_{51}$, and $r_{61} = -r_{22}$, so the coefficients are as follows.

$$r_{ij} = \begin{pmatrix} 0 & -r_{22} & r_{13} \\ 0 & r_{22} & r_{13} \\ 0 & 0 & r_{33} \\ 0 & r_{51} & 0 \\ r_{51} & 0 & 0 \\ -r_{22} & 0 & 0 \end{pmatrix} \tag{3-13}$$

The coefficients relevant to modulation are r_{22}, r_{13}, and r_{33}. Their values are $r_{22} = 3.4$, $r_{42} = 28$ $r_{13} = 8.6$, and $r_{33} = 30.8$, in units of 10^{-12} m/V. [5] If the applied electric field is along the z-direction, the equation of the index ellipsoid is simply given by

$$(\frac{1}{n_0^2}+r_{13}E_z)x^2 + (\frac{1}{n_0^2}+r_{13}E_z)y^2 + (\frac{1}{n_e^2}+r_{33}E_z)z^2 = 1 \qquad (3\text{-}14)$$

Eq. (3-14) can be again written as the form

$$\frac{x^2}{n_x^2}+\frac{y^2}{n_y^2}+\frac{z^2}{n_z^2}=1$$

Here $n_x = n_y \cong n_o - n_o^3 r_{13}E_z/2$, $n_z \cong n_e - n_e^3 r_{33}E_z/2$. In contrast to the usual anisotropic material, the refractive index is now further influenced by the applied field E_z.

In fact, LiNbO$_3$ is a popular material for the waveguide-type Mach-Zehnder modulator used in optical communication. The refractive index of bulk LiNbO$_3$ is as follows. [1]

$$n_0 = 2.195 + \frac{0.037}{\lambda^2}$$

$$n_e = 2.122 + \frac{0.031}{\lambda^2}$$

where the wavelength λ is in the unit of μm.

Fig. 3-9 shows an example of a waveguide made on x-cut LiNbO$_3$. The waveguide and electrodes are arranged such that the applied field is along the z-direction. The waveguide is along the y-direction. The light propagating in the waveguide is also polarized with the electric field along the z-direction.

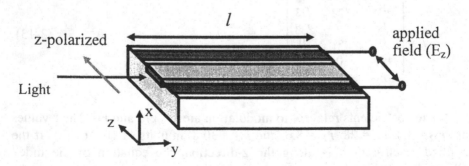

Figure 3-9. A x-cut LiNbO$_3$ with refractive index of the waveguide modulated by applied field. (blue color: waveguide, brown color: electrodes)

As the light propagates in the waveguide for a distance l, the electric field of the light takes on the phase modulated by the applied field E_z.

$$E = \hat{z}E_o e^{jkn_z y} = \hat{z}E_o e^{jn_e l - jn_e^3 r_{33} E_z l/2} = \hat{z}E_o e^{jn_e l} e^{-jn_e^3 r_{33} E_z l/2} \quad (3\text{-}15)$$

The phase term $\phi(E_z) = - n_e^3 r_{33} E_z \, l/2$ is a function of the applied field. When the fields applied to both waveguides of the Mach-Zehnder modulator are the same, the light waves passing through the two waveguides are in phase, as shown in Fig. 3-10 (a). When the two fields applied to both waveguides are properly different, the light waves passing through the two waveguides could be out of phase, as shown in Fig. 3-10 (b).

(a)

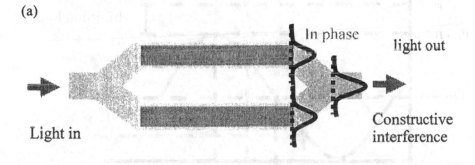

(b)

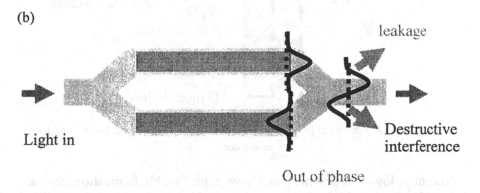

Figure 3-10. The output light in the two waveguides are (a) in phase; (b) out of phases.

Therefore, the output light is a function of the applied field or the voltage applied to the electrodes patterned next to the waveguide, as shown in Fig. 3-11.

Typical characteristics of a Mach-Zehnder modulator used in the optical communication are summarized as follows.

- The material commonly used is $LiNbO_3$.
- The modulation depth is usually better than 20 dB.
- The modulation bandwidth may reach 75 GHz. [6]
- Because the light has to be coupled to the input waveguide, there is coupling loss. In addition, the light divided to path 1 and path 2 experiences loss at the Y-junction. Therefore, the insertion loss is inevitable and is typically $\geq$ 4 dB.
- The light power that can be handled by the modulator is up to 200 mW.
- Because there is almost no carrier variation during modulation, the induced chirp is negligible.
- V_π is a few volts, depending on bandwidth.

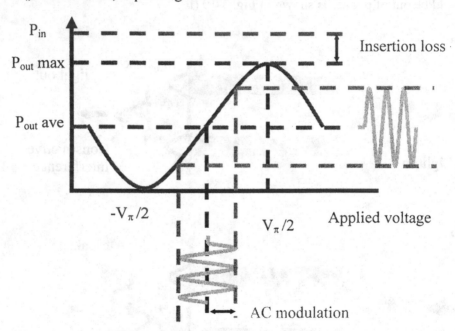

Figure 3-11. The output light varies with the applied voltage in the Mach-Zehnder modulator.

Recently, low-cost polymers that have more flexible fabrication steps are also used. [7] Some exhibit electro-optic coefficients as large as 55×10^{-12} m/V. With its high E-O coefficient, a polymer EO modulator can achieve a modulation bandwidth larger than 60 GHz and low operation voltage (V_π) of approximately 4.5 volts for an electrode length of 2 cm. [8] The polymer can be deposited on Si wafer, enabling the monolithic integration of E-O modulator with Si-based electronics.

3.2   Electro-Absorption (EA) Modulator

Another recently popular scheme of external modulation in optical communications is using electro-absorption (EA) modulation. Its operating principle may be understood from the reverse operation of light emitting diodes. As shown in Fig.3-12, when the p-n junction in an LED is reverse-biased, it absorbs light. Its depletion region extends as the p-n junction is strongly reverse-biased. Therefore, at zero bias, the absorption is weak. Under strong reverse bias, the absorption is strong. As a result, the light intensity passing through the p-n junction is modulated by the reverse-biased voltage.

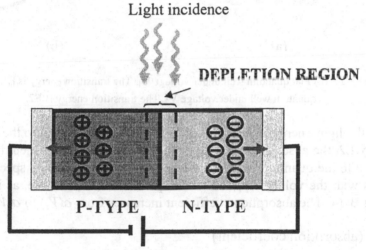

Figure 3-12. A reverse-biased p-n junction in LED becomes light absorption.

To enhance the effect of electro-absorption, a quantum-well structure is usually used. The quantized energy level of the quantum well is changed when a different voltage is applied, leading to a shift of the transition energy, called the Stark shift. The situation is illustrated in Fig.3-13(a) and (b). Under a certain bias voltage V_1, the quantum well has a flat bottom. (Fig. 3-13a). The quantized energy level can be calculated using the Schrödinger equation. Its energy levels in the conduction band and the valence band are schematically shown in the figure. The quantization of energy levels leads to the transition at energy E1. As the bias voltage is changed, the quantum well is tilted, as shown in Fig. 3-13 (b). The quantized energy level can be evaluated again using the Schrödinger equation. Because of the tilted quantum well, the energy levels in the conduction band and the valence band are changed, as schematically shown in Fig.3-13 (b). The transition energy in this case is reduced to E2, E2 < E1.

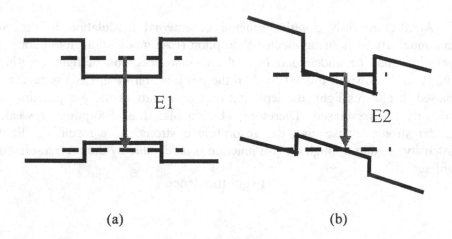

(a)   (b)

Figure 3-13. (a) Flat quantum well under voltage V_1. The transition energy is E1. (b) Tilted quantum well under voltage V_2. The transition energy is E2.

If the light energy incident on the EA modulator is equal to the transition energy E2, the absorption of the quantum well is increased when the voltage applied to the quantum well is changed to V_2. The absorption spectrum thus varies with the voltage applied to the quantum-well structure, as illustrated in Fig. 3-14. The absorption coefficient increases from $\alpha(V_1)$ to $\alpha(V_2)$.

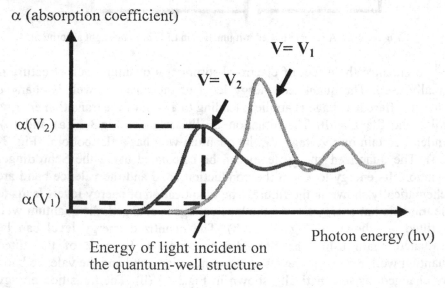

Figure 3-14. Absorption spectrum varies with the voltage applied to the quantum-well structure.

In actual applications, multiple quantum wells are possibly used to enhance the absorption effect. Fig. 3-15 shows a quantum-well EA modulator. The quantum wells are sandwiched between the two electrodes. A voltage V is applied between the two electrodes. If the modulator has a length L, the output power of the light is related to the input power by the formula

$$P_{out} = e^{-\alpha(V)L} P_{in} \qquad (3\text{-}16)$$

As $\alpha(V)$ increases, the output power decreases. Therefore, the output power is modulated by the applied voltage V.

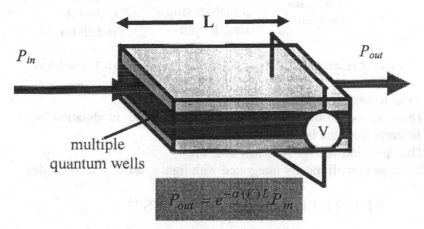

Figure 3-15. Quantum-well EA modulator.

Because the EA modulator is made of semiconductor material, it can be monolithically integrated with a laser diode. The monolithic integration has the advantages of reduced size and less insertion loss. A possible integration configuration is illustrated in Fig. 3-16. The quantum-well structure in the EA modulator must be different from that in the laser-diode section. In addition, there must be sufficient electrical insulation between the electrode of the laser diode and the electrode of the EA modulator.

The general characteristics of an EA modulator are as follows.
- The material is semiconductor quantum wells.
- The modulation depth can exceed 10 dB.
- The modulation bandwidth can be as high as 40 GHz. [9, 10]
- The insertion loss can approach zero if it is monolithically integrated with a laser diode. Even if it is not integrated with a laser diode, its insertion loss is less than the E-O Mach-Zehnder modulator because it has no Y-junction loss.

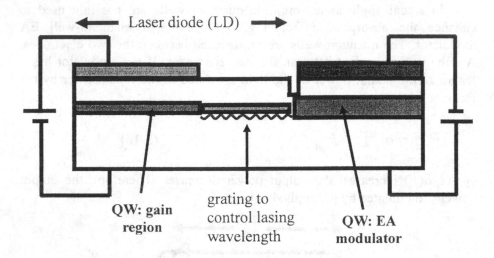

Figure 3-16. Monolithic integration of a laser diode and a QW EA modulator.

- It can handle approximately 1mW of power.
- There is negligible induced chirp because the modulation region is different from the laser region.
- The operation voltage is as low as 2 V.
- It can be monolithically integrated with light sources -- laser diodes.

4.   SHORT-PULSE TECHNIQUES [11]

In digital communication systems using optical fibers, the optical signals are coded with pulses. A train of the digital signals represented by optical pulses is illustrated in Fig. 3-17. The pulses are separated by a period T. The pulse repetition rate is thus equal to 1/T, which is also the bit rate of the communication system. Shown in Fig. 3-17 are the digital signals of "1 1 0 1 0 1". At each pulse position the presence of a pulse represents a "1". The absence of a pulse at the pulse position represents a "0". In order to clearly distinguish each "1" bit, the pulses should not overlap. Therefore, the pulse duration $\delta\tau$ must be less than the period T. Digital communications have the advantage of tolerating interference from background noise to a greater extent than analog communications.

The key issue for digital communications is the generation of optical pulses. Because semiconductor lasers are the most common light sources for optical communications, we will introduce the two major techniques used for short-pulse generation from semiconductor lasers: gain switching and mode-locking.

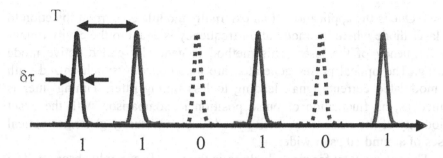

Figure 3-17. A train of the digital signals represented by optical pulses. The red solid line means an optical pulse at that temporal position for the "1" bit. The red dashed line means "no" optical pulse at that temporal position for the "0" bit.

4.1　Gain Switching

Gain switching is a simple technique that can generate optical pulses from semiconductor lasers. A large pulse of short duration can be generated utilizing the transient behavior of semiconductor lasers, where the gain of the semiconductor lasers suddenly increases to a level above the loss. This situation is illustrated in Fig 3-18. The sudden increase of gain can be achieved by the injection of a large current pulse. The optical pulses generated by this technique typically have a pulse width of not less than 10 ps.

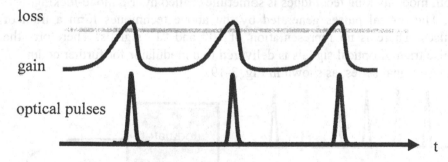

Figure 3-18. The time variation of gain in gain switching of semiconductor lasers. Optical pulses are generated approximately at the places of gain peaks.

4.2　Mode Locking

Another popular technique for short pulse generation is mode-locking. By locking the phases of all resonant modes, coherent interference of optical waves is achieved after each periodical time, so short pulses are generated periodically. Two common ways are used to lock the phases of the laser modes. One is the application of an externally modulated current injection to the laser diode where the modulation frequency is equal to the cavity round-trip frequency of the laser. This method is generally called active mode locking. The optical pulses generated are automatically synchronized with the modulated current signal, leading to less timing jitter. Timing jitter is defined as the fluctuation of pulse position in comparison with the exact periodicity of the pulse train. Active mode-locking usually generates optical pulses of around 10 ps or wider.

The second way for mode-locking is the use of saturable absorber. The saturable absorber is a material whose absorption is easily saturated by the light intensity. When the laser cavity has the saturable absorber, the portion of light with strong intensity experiences less absorption than the portion of light with weak intensity. Thus optical pulses are favorably supported in the laser cavity. Because the optical pulses are automatically generated in the laser cavity, no additional modulation on the current injection is necessary. On the other hand, the optical pulses are generated due to the saturation of absorption, so the saturation behavior usually varies with each optical pulse. As a result, the saturable absorber using saturable absorbers has a larger timing jitter than the active mode-locking. However, passive mode-locking may generate optical pulses of less than 1 ps, much shorter than the actively mode-locked pulses. To receive the advantages of short pulses and small timing jitter, both active mode-locking and passive mode-locking can be simultaneously applied to the semiconductor lasers. The simultaneous use of both mode-locking techniques is sometimes called hybrid mode-locking.

The optical pulses generated by the above techniques form a train of pulses. There is no representation of "0" and "1" signals. Therefore, the pulse train of optical signals is delivered to a modulator for further coding of binary signal series, as shown in Fig. 3-19.

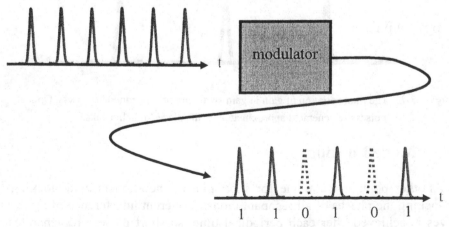

Figure 3-19. A pulse train is coded to "1 1 1 0 1 0 1" binary codes by a modulator.

NOTES

1. For mode information on modulation of laser diodes, readers are referred to Ref. 1.
2. For detailed explanation of chirp, readers are referred to Ref. 2. To understand the variation of index with injection current in laser diodes, please refer to Ref. 3.
3. Ref. 11 gives detailed explanation on the techniques for short-pulse generation.

REFERENCES

1. Yariv, Amnon, *Optical Electronics in Modern Communications*. 5/ed, Oxford University Press, 1997.
2. Siegman, A. E., *Lasers*. University Press, 1986.
3. Agrawal, Govind P. and Dutta, Niloy K., *Semiconductor Lasers*. 2/ed, Van Nostrand Reinhold, 1993.
4. Yariv, Amnon and Yeh, Pochi, *Optical Waves in Crystals*. John Wiley & Sons, 1984.
5. Boyd, G. D., Miller, R. C., Nassau, K., Bond, W. I., and Savage, A., LiNbO₃: an efficient phase matchable nonlinear optical material. Appl. Phys. Lett. 1964; 5:234-236.
6. Noguchi, K., Miyazawa, H., and Mitomi, O., 75-GHz broadband Ti:LiNbO₃ optical modulator with ridge structure. Electron. Lett. 1994; 30: 949-951.
7. Steier, W. H., Chen, A., Lee, S.-S, Garner, S., Zhang, H., Chuyanov, V., Dalton, L. R., Wang, F., Ren, A. S., Zhang, C., Todorova, G., Harper, A., Fetterman, H. R., Chen, D., Udupa, A., Bhattacharya, D., and Tsap, S., Polymer electro-optic devices for integrated optics. Chemical Physics 1999; 245: 487-506.
8. Chen, D., Fetterman, H. R., Chen, A., Steier, W. H., Dalton, L. R., Wang, W., and Shi, Y., Demonstration of 110 GHz electrooptic polymer modulators. Appl. Phys. Lett. 1997; 70: 3335-3337.
9. Ouagazzaden, A., Lentz, C. W., Mason, T. G. B., Glogovsky, K. G., Reynolds, C. L., Przybylek, G. J., Leibenguth, R. E., Kercher, T. L., Boardman, J. W., Rader, M. T., Geary, J. M., Walters, F. S., Peticolas, L. J., Freund, J. M., Chu, S. N. G., Sirenko, A., Jurchenko, R. J., Hybertsen, M. S., and Ketelsen, L. J. P., 40 Gb/s tandem electro-absorption modulator. Optical Fiber Communication Conference 2001, Anaheim, CA, postdeadline paper PD14.
10. Chou. H. F., Chiu, Y. J., and Bowers, J. E., 40GHz optical pulse generation using traveling-wave electroabsorption modulator. Optical Fiber Conference 2002. Technical Digest, Paper WV2, pp. 338-339.
11. Verdeyen, J. T., *Laser Electronics*. 3/ed, Prentice Hall, 1995.

Chapter 4

OPTICAL AMPLIFIERS

1.   INTRODUCTION

Optical communication systems are now widely spread over the entire earth. The world-wide link of optical fibers for signal transfer is more than 10,000 km. Even though the loss of optical fibers is as low as 0.2 dB/km, the loss over a long distance is still substantial. Therefore, the optical signals must be amplified after they propagate in the fiber for significant distances. The traditional way to amplify the optical signal is to convert the optical signal back to an electronic signal first, then amplify the signal in the electronic domain. The amplified electronic signal is then converted back into an optical signal again and delivered into the optical fiber. The block diagram of process is shown in Fig. 4-1.

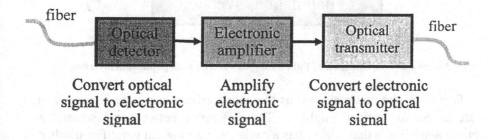

Figure 4-1. A conventional way to amplify the optical signal.

After 1990, optical amplifiers become popular to directly amplify the optical signals propagating in the optical fiber. The conversion of optical signals to electronic signals and vice versa is therefore no longer necessary. Nowadays, the erbium-doped fiber amplifier (EDFA) is the most common optical amplifier used in the optical communication systems. Semiconductor optical amplifiers (SOA) and Raman amplifiers have also attracted significant attention and recently increased in popularity. These three types of optical amplifiers will be introduced in this chapter.

2.   PRINCIPLE OF OPTICAL AMPLIFIERS

The basic function of an optical amplifier as their name implies is to amplify optical signals. In Chapter 2, we discussed the principle of lasers. There we mentioned that stimulated emission is a process which produces light amplification. Thus stimulated emission takes place in an optical amplifier. To make stimulated emission occur, population inversion of an atomic system is necessary. The material with population inversion will have a gain coefficient given by Eq. (2-21). As shown in Fig. 4-2, the light passing through such a material is amplified according to the following formula (Eq. (2-22)).

$$I_{out} = I_{in}e^{\gamma z}$$

where γ is the gain coefficient given by Eq. (2-21).

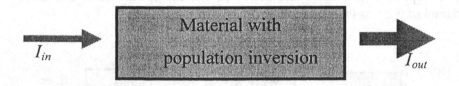

I_{in} → Material with population inversion → I_{out}

Figure 4-2. Amplification of light through a material with population inversion.

Therefore, the material used as the gain medium in a laser is exactly the material for an optical amplifier. The difference between a laser and an optical amplifier is that a laser has a cavity, but an optical amplifier needs no cavity. Fig. 4-3 clearly shows the difference among a usual light source, an optical amplifier, and a laser. In a usual light source, there is only spontaneous emission. In contrast, stimulated emission occurs in both an optical amplifier and a laser. However, there is no cavity for an optical amplifier, so light is amplified only through a single pass. In a laser, stimulated emission occurs. In addition, the light bounces forward and backward in the cavity, so light is amplified for each of the many trips through the cavity.

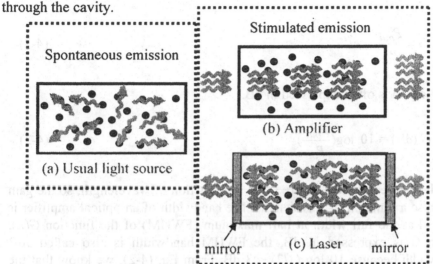

Figure 4-3. (a) In a usual light source, there is only spontaneous emission. (b) In an optical amplifier, stimulated emission occurs, but there is no cavity. Light is amplified through a single pass. (c) In a laser, stimulated emission occurs and the light bounces forward and backward in the cavity, so light amplification occurs for each of the many trips through the cavity.

To maintain the gain material at population inversion, an excitation system is also required. In summary, an optical amplifier consists of two major parts: (1) gain material for light amplification; (2) excitation system to maintain the gain material in population inversion.

2.1 Power Amplification

Eq. (2-22) describes the amplification of light intensity. However, the power level is what we are concerned with in optical communications. Because power is equal to the product of the light intensity and the cross section of the light beam, the amplification of light power still follows the same equation as Eq.(2-22).

$$P_{out} = P_{in}e^{\gamma z} \tag{4-1}$$

If the amplifier has a length of L, then $P_{out} = P_{in}e^{\gamma L}$. The gain of an optical amplifier is defined as the ratio of the output power to the input power.

$$G = \frac{P_{out}}{P_{in}} = e^{\gamma L} \qquad\qquad (4\text{-}2)$$

The gain is often expressed in dB.

$$G\,(dB) = 10\,\log(\frac{P_{out}}{P_{in}}) \qquad\qquad (4\text{-}3)$$

The gain coefficient in Eq. (4-2) is a function of wavelength, so the gain G is also a function of wavelength. The bandwidth of an optical amplifier is defined as the full width at half maximum (FWHM) of the function $G(\lambda)$. When G is expressed in dB, the FWHM bandwidth is also called 3dB bandwidth because 10 log (1/2) ~ 3 dB. From Eq. (4-2), we know that the FWHM bandwidth of the amplifier is smaller than the FWHM bandwidth of the gain coefficient $\gamma(\lambda)$.

2.2   Gain Saturation

Eq. (2-22) or Eq.(4-2) is derived from Eq. (2-20). In that derivation, the coefficient γ is assumed to be constant over the propagation distance z. If the saturation effect is taken into account, the coefficient γ is actually a function of intensity, as described in Eq. (2-30). Again, since we are concerned with power, Eq.(2-30) is replaced by

$$\gamma = \frac{\gamma_0}{1 + P/P_s} \qquad\qquad (4\text{-}4)$$

where P_s is the saturation power and is equal to the product of the saturation intensity and the cross section of the light beam. Also, Eq. (2-20) is replaced by

$$\frac{dP}{dz} = \gamma P = \frac{\gamma_0 P}{1 + P/P_s} \qquad\qquad (4\text{-}5)$$

Again, defining $P_{in} = P(0)$ and $P_{out} = P(L) = G\,P_{in}$ for the amplifier with a length L, we are able to solve the saturation gain, G, from Eq. (4-5). It is in an implicit form. [1,2]

$$G = G_0 \exp\left(-\frac{P_{out} - P_{in}}{P_s}\right) = G_0 \exp\left(-\frac{G-1}{G}\frac{P_{out}}{P_s}\right) \tag{4-6}$$

where G_0 is the unsaturated gain, $G_0 = e^{\gamma_0 L}$. From Eq. (4-6), we can see that G is approximately equal to G_0 when the output power is small. As the output power increases, G decreases. The output power for which the gain is reduced to one half of the unsaturated gain is defined as the output saturation power, P_{out}^s. Setting $G = G_0 / 2$ in Eq. (4-6), we can solve the output saturation power. [1,2]

$$P_{out}^s = \frac{G_0 \ln 2}{G_0 - 2} P_s \tag{4-7}$$

In practice, an unsaturated gain of more than 20 dB is common for an optical amplifier. Then $G_0 > 100$, so $P_{out}^s \approx (\ln 2)P_s \sim 0.7 P_s$.

3. ERBIUM-DOPED FIBER AMPLIFIER (EDFA)

3.1 Gain Material: Erbium-Doped Fiber

The erbium-doped fiber amplifier (EDFA) has the gain material made of an optical fiber doped with erbium ions (Er^{3+}). The erbium-doped fiber has the energy levels shown in Fig. 4-4. [3-5] From those energy levels, we know that the erbium-doped fiber has the absorption lines at 532 nm, 670 nm, 800nm, 980nm, and 1480 nm. Thus the incidence of the light at wavelength 532 nm causes the transition from the energy level $^4I_{15/2}$ to the energy level $^4S_{3/2}$. The 670 nm light causes the transition from the energy level $^4I_{15/2}$ to the energy level $^4F_{9/2}$. Similarly, the light at other wavelengths causes the corresponding transitions shown in Fig. 4-4. The most popularly used absorption lines are 980nm and 1480 nm. The symbols $^4I_{15/2}$, $^4I_{13/2}$, $\cdots$ represent the energy levels according to the definition in quantum mechanics.

As electrons are pumped to the high energy levels, they will have radiative transition to emit light, as shown in Fig. 4-4. For example, the transition from the high-energy level $^4I_{13/2}$ to the low energy level $^4I_{15/2}$ will emit light at the wavelength around 1540 nm.

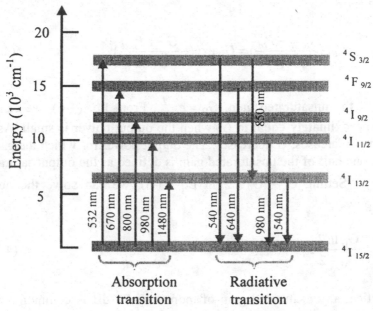

Figure 4-4. Energy levels in Er-doped fiber. (The unit of energy is given by the inverse of wavelength, $1/\lambda(cm)$.)

3.2   Working as a Four-Level Atomic System

The gain material of the Er-doped fiber can be treated simply as a four-level atomic system shown in Fig. 4-5.[6] The absorption of light at the pumping wavelength causes electrons to transit to the E3 high energy state. Electrons at the E3 state relax quickly to the E2 state, then transit to the E1 state by stimulated emission due to the input light with the energy equal to (E2-E1). Electrons in the E1 state also quickly relax to ground state E0. States E2 and E1 are always under population inversion.

If the pumping wavelength is 980 nm, then the E3 state represents the energy level $^4I_{11/2}$ in the Er-doped fiber. The $^4I_{13/2}$ energy level actually consists of many closely spaced energy levels, forming an energy band. If the pumping wavelength is 1480 nm, the E3 state represents the top level of the $^4I_{13/2}$ energy band. Electrons in those levels could quickly relax to the bottom levels of the $^4I_{13/2}$ energy band. Thus the E2 state represents the bottom level of the $^4I_{13/2}$ energy band. The $^4I_{15/2}$ energy level also consists of many closely spaced energy levels that form an energy band. The top levels and the bottom levels in this band are represented as the E1 state and the E0 state in Fig. 4-5, respectively. Stimulated emission therefore corresponds to the wavelength around 1540 nm.

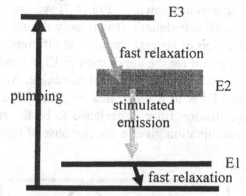

Figure 4-5. A four-level system representing the energy levels of gain medium of Er-doped fiber.

The pumping sources are usually laser diodes that emit light at 980 nm or 1480 nm.[7] The light from the pumping source is delivered to the Er-doped optical fiber using a fiber coupler. A schematic of the configuration containing the EDFA and the pumping source is shown in Fig. 4-6. The Er-doped fiber is usually a few meters.

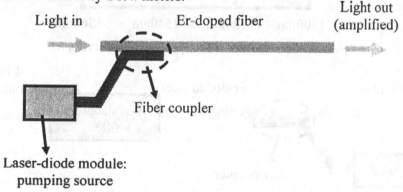

Figure 4-6. A schematic of the Er-doped fiber amplifier and the pumping source.

3.3   Gain and Amplification

Because $^4I_{13/2}$ and $^4I_{15/2}$ are energy bands, the transition from $^4I_{13/2}$ to $^4I_{15/2}$ involves the spectrum covering C-band (1525-1565nm) and L-band (1570-1610 nm). Typically the gain of EDFA in the C-band is much larger than in the L-band. The gain spectrum with the 980 nm pumping source is shown in Fig. 4-7 and represents only a typical value. The actual gain value depends on the doping concentration of Er ions in the fiber and the pumping power of delivered into the Er-doped fiber.

From the gain spectrum shown in Fig. 4-7, we see the gain spectrum varies significantly with wavelength. It is usually desirable to have an equal gain for the amplification of several channels at different wavelengths. Gain flattening is thus important for the application of EDFA in the WDM system. Fig. 4-8 illustrates the configuration of gain flattening. An equalization filter is inserted after the EDFA to achieve this purpose. [8] Ideally, the spectrum response of the equalization filter is designed to be the reverse of the gain spectrum, so their combination gives a flat response of light amplification, as shown in Fig. 4-9.

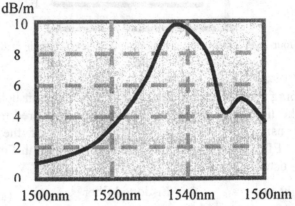

Figure 4-7. Gain spectrum of Er-doped fiber amplifier (EDFA).

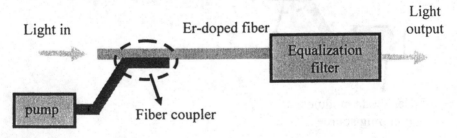

Figure 4-8. Gain flattening technique.

In the amplifier, not only is the signal amplified, but the spontaneous emission is amplified as well. [9] The amplified spontaneous emission (ASE) is the major source of noise for an optical amplifier. EDFA also exhibits ASE noise. ASE noise varies with the signal level, but not by much. [10] The variation of the ASE noise with the power of input light is shown in Figure 4-10.

Like usual gain media, EDFA also has gain saturation, which causes the gain to decrease with the optical power. The variation of gain with the power of input light is also shown in Fig. 4-10.

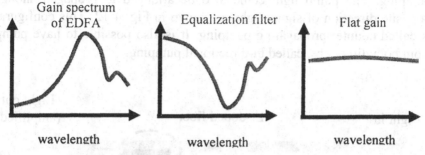

Figure 4-9. Concept of gain flattening technique.

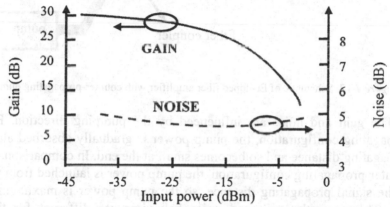

Figure 4-10. Gain saturation and variation of noise with input power.

Because EDFA usually has a length of more than one meter, the power of pumping power is absorbed gradually along the propagating distance in the Er-doped fiber. Hence the pump power gradually decreases along the propagating distance. On the other hand, the amplification happens for the entire propagating distance in EDFA. Typical variation of pump power and signal power along the propagating distance is shown in Fig. 4-11. [10]

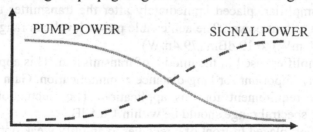

Propagation distance of fiber amplifier

Figure 4-11. variation of pump power and signal power along the propagating distance.

In Fig. 4-6, the pump light propagates along the same direction as the signal light. This configuration is called forward pumping or co-propagating pumping. The pump light could also be arranged to propagate along the opposite direction of signal light, as shown in Fig. 4-12. This configuration is called counter-propagating pumping. It is also possible to have pumping from both directions, called bi-directional pumping.

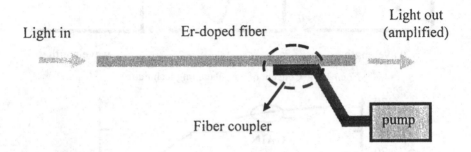

Figure 4-12. Schematic of Er-doped fiber amplifier with counter-propagating pumping.

The gain and noise are influenced by the pumping direction. For co-propagating configuration, the pump power is gradually absorbed along the propagating distance and so becomes small at the end. In comparison, for the counter-propagating configuration, the pump power is launched from the end of the signal propagating distance, so the pump power is maximum at the end. Therefore, the gain and noise behaviors are different for the two configurations.

3.4   Applications of Er-Doped Fiber Amplifiers

The main application of Er-doped fiber amplifiers in optical communications is to amplify the optical signal. Such application can be divided to three types.

(1) Booster amplifier: placed immediately after the transmitter to provide maximum output power. The achievable power level can range from 16 dBm (39.8 mW) to 19 dBm (79.4mW).

(2) In-line amplifier: used in the middle of transmission. This application is particularly important for long-distance communication. Gain flatness is the major requirement for this application. The fluctuation of gain across the spectral range should be within ± 0.5dB.

(3) Preamplifier: placed in front of a receiver to amplify weak optical signal after long transmission through the fiber. The important features for this application are low noise, high gain, and high sensitivity.

3.5 Characteristics of Er-Doped Fiber Amplifiers

The characteristics of Er-doped fiber amplifiers being used as booster amplifiers, in-line amplifiers, and preamplifiers are slightly different. When used as booster amplifiers and in-line amplifiers, several channels of different wavelengths may be amplified simultaneously. When used as a preamplifier, only one channel is amplified because each receiver takes care of only one optical channel. The characteristics are separately summarized as follows.

- **Booster amplifiers**
  small signal gain .... 30 dB (typical)
  gain per channel .... 10-20 dB
  input level (total) .... -5 to +5 dBm
  input level (32 channels) .... -20 to -10 dBm (per channel)
  input level (16 channels) .... -17 to -7 dBm (per channel)
  saturation power (spectral range: 1528-1563nm)
  .... 16-19 dBm
  saturation power (spectral range: 1568-1603nm)
  .... 16-18 dBm
  gain flatness .... $< \pm 1.0$dB

- **In-line Amplifiers**
  small signal gain .... 35 dB (typical)
  gain per channel .... 20-30 dB
  composite input level .... -20 to -5 dBm
  input level (32 channels) .... -35 to -20 dBm (per channel)
  input level (16 channels) .... -32 to -17 dBm (per channel)
  saturation power .... 14-18 dBm
  gain flatness .... ± 0.5dB (typ.), ± 0.8dB (max.)

- **Pre-amplifier**
  small signal gain .... 35 dB (typical)
  typical gain constant .... 25-35 dB
  composite input level .... -40 to -20 dBm
  composite output level .... -5 to 5 dBm
  gain flatness .... $< \pm 1.0$dB

3.6 Other Types of Fiber Amplifiers

The Er-doped fiber amplifiers are good for the spectral range in C-band (1525-1565nm) and L-band (1570-1610 nm). However, the usable spectral range of the optical fiber is from less than 1.3 μm to above 1.6 μm. Most of the region is not covered by the Er-doped fiber amplifiers. Therefore, other types of fiber amplifiers are developed for use in other spectral regions. They are introduced in the following.

1. Praseodymium (Pr) doped fiber amplifier: [11, 12] The fiber is doped with Pr ions. It has the absorption lines are at 1017 nm or 1047 nm, so the pumping sources could be InGaAs laser diodes at 1017 nm or Nd:YLF solid-state lasers at 1047 nm. The stimulated emission for amplification is at the 1300 nm band.

2. Neodymium (Nd) doped fiber amplifier: The fiber is doped with Nd ions. It is very similar to the gain medium of Nd: Glass laser. The absorption is around 795-810 nm, so high-power laser diodes at 795 or 810 nm could be used as the pumping sources. Nd-doped fiber amplifiers could provide amplification for the spectral range 1310 - 1360 nm.

3. Plastic fiber amplifier: The plastic fibers have good transmission at 570 nm and 670 nm. The Rhodamine B doped PMMA (polymethylmethyl-acrylate) fiber could have the gain window in 610 nm - 640 nm. Then the popular red-light laser diodes at 635 nm could be used for transmitters. However, plastic fibers still have much larger loss than glass fiber, so they may be useful only for short-distance communication.

4.   SEMICONDUCTOR OPTICAL AMPLIFIER (SOA)

Semiconductor optical amplifiers (SOAs) as the name suggests are made of semiconductors. In fact, semiconductor optical amplifiers are the gain media of semiconductor lasers. As the cavity of semiconductor lasers is removed, they become SOAs. It sounds quite easy. However, as explained in Section 2.3, the cavity of semiconductor lasers is formed by naturally cleaved facets. Those naturally cleaved facets automatically give a reflectivity of 0.3, which is sufficient for the oscillation of light. In other words, there are no additional elements added to form the cavity of semiconductor lasers. Therefore, to remove the cavity does not mean taking components away from the semiconductor lasers. Instead, we have to do extra work to make the naturally cleaved facets non-reflective. That is, we have to significantly reduce or eliminate the reflectivity of the cleaved facets. Therefore, the semiconductor optical amplifiers have very similar structures to the laser diodes except that the facets are modified to have very low reflection. With very low reflection at the cleaved facets, the chip of the laser diode works as an optical amplifier. Its function is schematically shown in Fig. 4-13.

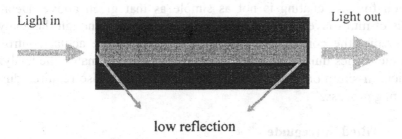

low reflection

Figure 4-13. Schematic function of an SOA.

4.1 Schemes to Reduce Facet Reflectivity

Several schemes are adapted to reduce the facet reflectivity. They are
(1) Anti-reflection (AR) coating
(2) Tilted waveguide
(3) Curved waveguide
(4) Tilted facet
(5) Window facet

4.1.1 AR-Coating

AR-coating is a common technique for reducing the reflectivity of an optical component. [13] In principle, for plane waves incident on the facet, the reflectivity can be reduced to zero by a single layer of dielectric coating. The condition is that the dielectric should have the refractive index equal to $n_{1/2}$ and its thickness equals $\lambda_{coating}/4$, where n is the refractive index of the substrate and $\lambda_{coating}$ is the wavelength in the coating layer. In reality, the light guided in the waveguide of the SOA is not composed of plane waves. In addition, it is difficult to control the coating material to have the refractive index and the thickness exactly equal to $n^{1/2}$ and $\lambda_{coating}/4$, respectively. Furthermore, the above condition is good only for a single wavelength. However, SOAs are often used for a bandwidth more than 30 nm. Therefore, the reduction of the reflectivity to zero for the working bandwidth of 30 nm is not possible. In practice, low reflectivities are obtained by in-situ monitoring the SOA characteristics during the coating process. 10^{-4} of reflectivity can be achieved for SOAs with a bandwidth of 20 nm. Because the refractive index of the SOA material is approximately 3.5, the refractive index of the coating dielectric is about 1.87. Dielectric materials like SiO and SiN are commonly used for single-layer AR coating.

Multi-layer coating usually gives better performance than single-layer coating, in particular for bandwidths exceeding 20 nm. Analysis for multi-layer coating is much more complicated than single-layer coating. The condition for AR coating is not as simple as that given above. Detailed analysis of multi-layer coating is given in Chapter 6. Although the analysis could predict the condition for required reflectivity, the actual control of coated dielectrics during coating process cannot exactly match the analyzed condition. In-situ monitoring the SOA characteristics is also required during the coating process.

4.1.2   Tilted Waveguide

The SOA with a tilted waveguide has its structure similar to the Fabry-Perot laser diode except that the waveguide is not normal to the cleaved facets. A schematic diagram of the top view of the device is shown in Fig. 4-14. The width of the waveguide is the same as the waveguide in the laser diode, about 2-5 μm. The waveguide is oriented at an angle θ from the normal of the cleaved facets. [14] An angle between 5° and 12° usually provides good results. [15] As shown in the figure, the reflected wave does not travel back to the waveguide. Instead, it is reflected toward a direction that is at an angle 2θ from the waveguide and then goes off the waveguide. The region outside the waveguide is not pumped and so it can absorb the reflected wave. Thus the reflected wave disappears after it travels off the waveguide for some distance. Because no wave is reflected back to the waveguide, light resonance between the two cleaved facets is eliminated.

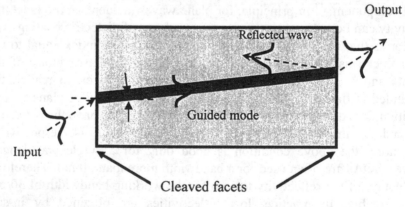

Figure 4-14. SOA with the tilted waveguide.

The above description is for the ideal case. In reality, the guided mode in the waveguide is composed of many plane waves that travel toward different directions. Most of the decomposed plane waves propagate at directions not very different from the waveguide direction. However, the propagation directions of some waves still deviate from the waveguide direction for an angle near θ. Those components can be directly reflected by the cleaved facets and then become part of the guided mode propagating backward. Therefore, there is some reflection. The reflection is a function of the angle θ and the waveguide width. For the angle of 7°, the equivalent reflectivity (defined as the ratio of intensity reflected back to the waveguide) could be about 0.2 %. Usually additional AR-coating is applied to further reduce the equivalent reflectivity. The AR-coating does not have to be very good in this case. Combining AR-coating of around 1 % and the tilted waveguide can easily achieve an equivalent reflectivity of less than 10^{-4}.

It is also clear that the transmitted wave is propagating at a direction deviating from the normal of cleaved facets for an angle much larger than q as a result of Snell's law. In addition, due to the large deviation from the facet normal, the beam shape is distorted. It has a crescent shape instead of an elliptical one.

4.1.3 Curved Waveguide

The SOA with a curved waveguide also has a structure similar to the Fabry-Perot laser diode except that the waveguide is curved so that it is not normal to the cleaved facets. The curved waveguide provides the flexibility that the waveguide could be oriented at two different angles from the cleaved facet. [16, 17] Usually, one side of the waveguide is still normal to the cleaved facet, while the other side is oriented at an angle θ from the normal of the cleaved facet, as shown in Fig. 4-15. An angle between 5° and 12° will usually provide good results. [15]

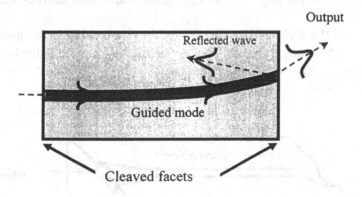

Figure 4-15. SOA with the curved waveguide.

The bending side of the waveguide has a similar function to the tilted waveguide. The reflected wave propagates toward a direction deviating from the waveguide for an angle of 2θ. Because the reflected wave does not go back into the waveguide, light resonance between the two cleaved facets does not happen. The equivalent reflectivity at the bending side is similar to the value provide by the tilted waveguide. The curved-waveguide SOA is convenient for use when one side of the cleaved facet still serves as the reflection mirror, e.g., in an external-cavity configuration of tunable semiconductor lasers.

The curved waveguide has some disadvantages. First, it introduces bending loss. [17] Second, the wave guiding may be bad if the waveguide is not well fabricated. For example, if the edge of the ridge waveguide is rough, then the guiding effect is not sufficient to make light propagate along the curved waveguide.

4.1.4   Tilted Facet

In this structure, the waveguide is still aligned normal to the cleaved facets like the case of laser diodes. However, the waveguide is stopped at an etched facet instead of the naturally cleaved facet. [18] The etched facet is at an angle θ from the naturally cleaved facet, as shown in Fig. 4-16. Similar to above cases, an angle between 5° and 12° will usually provide good results. A dry-etching technique is usually applied to fabricate the facet. As the wave propagates to the end of the waveguide, it is reflected by the etched facet. Because the waveguide is not normal to the cleaved facet, the reflected wave goes toward a direction that is at an angle 2θ from the waveguide. Again, because the reflected wave does not go back to the waveguide, no light bounces forward and backward in the waveguide. The equivalent reflectivity is a function of the angle θ, the waveguide width, and the roughness of the etched facet.

Because the etched facets are created by dry-etching techniques, they do not have to be along a specific crystal orientation. Therefore, the two cleaved facets can be aligned at different angles from the waveguide according to the desired purpose.

(a)

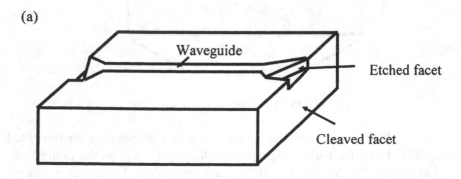

(b)

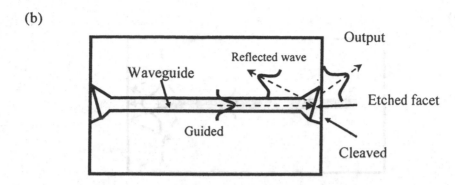

Figure 4-16. (a) The structure of SOA with the tilted facets. (b) The top view of the above structure.

4.1.5   Window Facet [19]

The structure of window facet incorporated into the SOA is shown in Fig. 4-17. In this structure, the waveguide is also aligned normal to the cleaved facet. However, the waveguide is not in direct contact with the cleaved facet. Instead, a region is introduced between the waveguide and the cleaved facet. This region is transparent and has no waveguide. Thus as the guiding mode emits into this region, it starts to spread out due to the diffraction of wave. As the mode is reflected by the cleaved facet, it spreads out even further. Only a very small portion of the wave goes back into the waveguide. Most of the wave travels in the region outside of the waveguide, which is not pumped and hence can absorb the reflected wave. Thus most of the reflected wave disappears after it travels for some distance. The window facet can result in a reflectivity as low as 10^{-4}. However, to make the window-facet structure, epitaxial regrowth is required, which could significantly complicate the fabrication process and increase the cost.

(a)

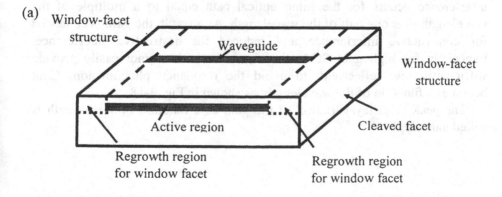

(b)

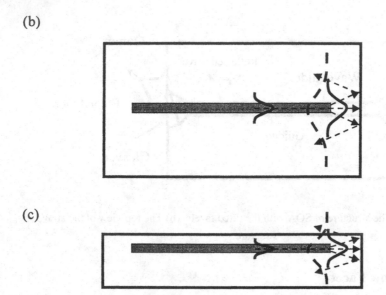

(c)

Figure 4-17. (a) The structure of SOA with the window facets. (b) The top view of the above structure. (c) The side view of the above structure.

4.2   Effect of Facet Reflectivity on the Gain of SOA

Because the reflectivity of the facets cannot be reduced to zero, it is important to know the effect of facet reflectivity on the gain of SOA. When the facet has reflection, some portion of light will bounce between the two facets leading to resonance. In particular, because the SOA has gain, the resonance effect is further enhanced since even a small amount of reflection will be amplified by the gain medium. Therefore, just like Fabry-Perot cavity, constructive interference happens when the round trip of the optical path between the two facets is equal to a multiple of the wavelength. Destructive interference occurs for the same optical path equal to a multiple of the wavelength plus one half of the wavelength. As a result, the gain is enhanced for constructive interference and reduced for destructive interference. Especially for high gain SOA, the above methods cannot easily provide sufficiently low reflectivity to avoid the resonance phenomenon. Gain becomes a function of the wavelength, as shown in Fig. 4-18.

The peak-to-valley variation of the gain as a function of wavelength is called gain ripple. [20]

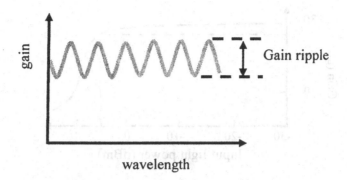

Figure 4-18. Variation of the gain with wavelength.

$$\Delta G = (\frac{1 + \sqrt{R_1 R_2} G}{1 - \sqrt{R_1 R_2} G})^2 \tag{4-8}$$

where G is the gain of the SOA chip, R_1 and R_2 are the reflectivity of the two facets. For an SOA with 30 dB gain, the reflectivity should be less than 1.2×10^{-5} in order to make the gain ripple less than 0.2dB. Thus AR-coating is usually not good enough for the low reflectivity. Several schemes mentioned above may be applied together to reduce the gain ripple to a satisfactory value.

4.3 Gain of an SOA

Similar to laser diodes, semiconductor optical amplifiers have a p-n junction and quantum wells. The band diagram of an SOA with quantum wells is similar to that shown in Fig. 2-33. When the device is under forward bias, electrons and holes are injected into the quantum wells. With sufficient carriers in quantum wells, population inversion can be achieved and light amplification is enabled. The gain coefficient is similar to that of laser diodes. For a gain coefficient of 100 cm^{-1} and a length of 700 μm, the gain of the SOA chip is about 30 dB according to Eq. (4-2). Gain saturation also occurs when the light power is large. A typical saturation behavior is shown in Fig. 4-19.

4.4 Characteristics of SOAs

The characteristics of semiconductor optical amplifiers are summarized in the following.

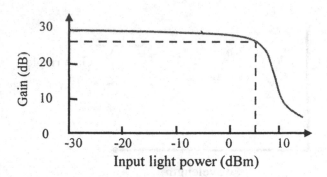

Figure 4-19. Gain saturation of SOA.

- They are made of semiconductor materials.
- They are fabricated using Semiconductor processing techniques.
- The working wavelength of the SOA is approximately given by wavelength (λ, in μm) = 1.24/E (eV), where E is the energy difference of the quantized level of quantum wells in the conduction band and the quantized level in the valence band.
- They are commonly fabricated from the following semiconductor materials:

Material	Bandgap (eV)	Wavelength (μm)
GaAs	1.42	0.87
InP	1.35	0.92
InGaAsP	0.73 - 1.35	0.9 - 1.7

- The gain bandwidth is 30 nm ~ 50 nm.
- The saturation power is about 13 dB (20 mW).
- The signal gain is about 10 dB ~ 30 dB.
- The input power level can be as low as -30 dBm (1 μW).
- There is usually a strong polarization dependence. The light of TE polarization usually has the gain about 5~7 dB larger than the TM-polarized light. Here the TE polarization is defined as the electrical field being parallel to the plane of p-n junction. The TM polarization is perpendicular to the TE polarization.
- The spontaneous emission is also amplified significantly since the gain coefficient of SOA is very large. The ASE noise is typically above 7 dB.

4.5   Comparison between SOA and EDFA

Both SOA and EDFA can be used to amplify the optical signal in the fiber-optic communication. They each have some fundamental differences which provide both advantages and disadvantages. A comparison between the two types of amplifiers is provided below.

SOA	*EDFA*
time constant 0.2 ns	time constant 10 ms
operation away from saturation	operation under saturation
power limit ~ 20 mW	power limit ~ 100 mW
more gain competition	less gain competition
more cross talk	less cross talk
polarization dependent	polarization insensitive
compact: integration on PCB	bulky: rack mount
large flat gain spectrum	narrow peak gain spectrum
many available gain spectra	only for C-band and L-band
gain bandwidth: 40 nm	gain bandwidth: 80 nm (C+L)
switchable	non-switchable
competitive cost	high-end price

4.6 Package of SOAs

Because the SOA has the structure similar to a LD, its dimension is very small. Its length is approximately between 500 μm to 1000 μm. The device width and height are about 300 μm and 100 μm, respectively. The light emission or amplification cross section is about 0.2 μm x 3 μm, where 0.2 μm is the thickness of active layer and 3 μm is the width of waveguide. The small size makes the light coupling between the SOA and the optical fiber very difficult. It is not possible for a typical customer to directly use an SOA chip in a fiber-optic communication system. Therefore, the SOA has to be packaged in a special way to facilitate its use in the fiber-optic communication system. A fiber-pigtailed package is a commonly used package for this application. It is schematically shown in Fig. 4-20.

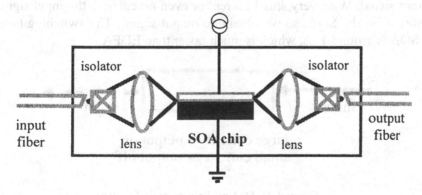

Figure 4-20. A fiber-pigtailed package of SOA.

Because the mode shape in the fiber and the waveguide of an SOA are not the same, a special type of lens may be required in order to provide optimal coupling efficiency of light between the SOA and the fiber. Sometimes the lens is directly made at the end of the fibers to simplify the procedure of alignment for coupling.

4.7   Applications of SOAs

In addition to amplification of optical signals, SOAs can also be used for other applications described in the following.

4.7.1   External-Cavity Tunable Semiconductor Laser

Because the facets of SOAs have very low reflectivity, it does not oscillate by itself. Because the SOA has gain however, if it is inserted into a cavity it will oscillate. If there is an optical component with the capability of wavelength selection in the cavity, it will oscillate at the selected wavelength. For example, if a diffraction grating is used, as shown in Fig. 2-39, the lasing wavelength is selected by the grating. By changing the orientation of the grating, the selected wavelength is tunable. The tuning of wavelength can also be achieved using a tunable fiber grating or a thin-film filter, which will be described later in Chapter 7.

4.7.2   Optical Switch

The gain of the SOA depends on the population, i.e., the injection current. If the injection current is very low, population inversion is not achieved. In this case the SOA will absorb light. This means that the SOA can function as an optical switch by changing the current level. This is shown in Fig. 4-21. With large current, the input signal passes through the SOA, so we obtain the output signal. With very small current or even no current, the input signal is absorbed by the SOA, so we obtain no output signal. The switching time of an SOA is around 1 ns, which is much faster than EDFA.

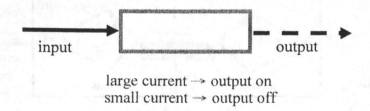

input   output

large current → output on
small current → output off

Figure 4-21. SOA works as an optical switch.

Several SOAs can be further combined to form a switch fabric. [21-23] Fig. 4-22 shows a 2 x 2 switch (two inputs and two outputs). Four SOAs are used in this configuration. There are named as (a)-1, (a)-2, (b)-1, and (b)-2, respectively, as shown in Fig. 4-22. There are two inputs, labeled A1 and A2, and two outputs, labeled B1 and B2. If (a)-1 and (b)-1 are injected with a large current, while (a)-2 and (b)-2 are injected with a small current, the input A1 is delivered to the output B1. If (a)-1 and (b)-2 are injected with a large current, while (a)-2 and (b)-1 are injected with a small current, the input A1 is delivered to the output B2. Other switching effects can be achieved by properly biasing the four SOAs.

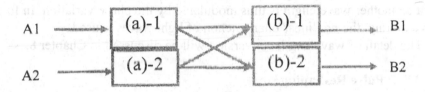

Figure 4-22. A 2 x 2 optical switch using four SOAs.

4.7.3   Wavelength Conversion

When two modes of different wavelengths exist simultaneously in the SOA, the two modes have strong competition for the available population inversion. This phenomenon is called gain competition. It can be used for wavelength conversion. This mechanism is schematically shown in Fig. 4-23. The light at wavelength λ1 has a modulation signal. The light at a second wavelength λ2 has no modulation. They are coupled into the same SOA simultaneously. Due to the gain competition, when the intensity of the light at wavelength λ1 is strong, the intensity of the light at wavelength λ2 is weak. Similarly, when the intensity of the light at wavelength λ1 is weak, the intensity of the light at wavelength λ2 is strong. Therefore, the modulation signal at wavelength λ1 is converted to wavelength λ2, but with an inverted shape.

Wavelength conversion using SOAs can also be achieved using other mechanisms like cross phase modulation and four-wave mixing. In cross phase modulation, two SOAs are arranged in the Mach-Zehnder interferometer. Because the SOA is made of semiconductor material, its refractive index greatly varies with the carrier density. When the intensity of light is modulated, it will cause the carrier density to change and so the

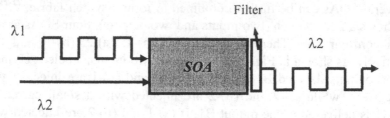

Figure 4-23. Wavelength conversion due to gain competition between wavelengths λ1 and λ2.

refractive index varies according to the modulation of light. In the Mach-Zehnder interferometer, the light passing through one of the paths has its phase modulated by the light with the modulation signal. The intensity of the light at another wavelength is thus modulated by the phase variation. In four-wave mixing, the nonlinear phenomenon of light waves is used.

The detail of wavelength conversion will be discussed in Chapter 8.

4.7.4   Pulse Reshaping

In digital communication, "1" is represented by a nonzero pulse, while "0" is represented by zero intensity of light. During the propagation of light signals, some background noise is added to the train of signals. Sometimes the background is large and can confuse the discrimination of the "0" level, causing an error detection of the "0" signal. The SOA can be used to reduce the background noise using the absorption saturation. The concept is illustrated in Fig. 4-24. The absorption saturation causes the absorption to decrease with the intensity. Thus the strong portion of the pulse will pass through the SOA, while the low level of noise intensity is absorbed. In this way, the train of pulse signals is reshaped with a much cleaner background, so erroneous decision of "0"level is reduced.

Figure 4-24. Pulse reshaping using a SOA to reduce the noise level of background.

5. RAMAN AMPLIFIER

In 1990s, Er-doped fiber amplifiers (EDFAs) were the major type of optical amplifiers used in the optical communication systems. Due to the growth of data communications however, the bandwidth provided by EDFAs became insufficient. Therefore, there was a strong desire to expand the bandwidth of the optical amplifiers. Raman amplifiers have attracted much attention in recent years because of their capability to provide gain at any wavelength.

5.1 Raman Scattering

Raman amplifiers are associated with Raman scattering, a phenomenon discovered by Indian physicist Sir C. V. Raman in 1928. [24] Raman scattering is a phenomenon of nonlinear optics. When light propagates through a medium, another beam of light at a downshifted frequency is also generated. The amount of downshifted frequency is equal to the frequency of the vibrational mode of atoms in the medium. As we explained in Chapter 1, when light is incident on a material, the electric field will cause polarization, .i.e., the separation of positive and negative charges. Because the electric field of light is oscillating, the dipoles formed by the separated positive and negative charges are also oscillating. At the same time, the molecules in the material have their own vibrations with the vibrational frequency determined by the molecular structure. As a result, the dipole oscillation induced by the electric field and the molecule vibration exist simultaneously, making the polarization have the term that is simultaneously influenced by the incident light and the molecule vibration. The polarization then further generates additional electric fields that are associated with the incident light and the molecular vibration. Because the polarization has an oscillating frequency different from the frequency of the incident light by an amount of the molecule's vibrational frequency, the Raman scattering causes another beam of light at the downshifted frequency. In fact, a similar effect could also generate light at an upshifted frequency. The light with downshifted frequency is called a Stokes wave, while the light with upshifted frequency is called an anti-Stokes wave. The incident light is called the pump wave.

The vibration of a molecule in the material can be formulated as quantization of phonon. Just like photon representing the quantization of EM wave, phonon is the quantization of acoustic wave. [25] Most of solids have two branches of the vibration energy (or phonon energy) vs. wave vector of the acoustic wave. The top branch represents phonons with the high energy and corresponds to a high vibrational frequency. Such phonons are called optical phonons. The bottom branch represents phonons with the low energy and corresponds to a low vibrational frequency. Such phonons are called acoustic phonons. Raman scattering can be treated as the interaction of photon and phonon. When the incident photon gives away part of its energy to an optical phonon, another photon of the Stokes wave with a downshifted frequency is generated. It is also possible that the photon interacts with the acoustic phonon. This process is called Brillouin scattering. [26]

5.2   Stimulated Raman Scattering

Raman scattering can be simulated by energy levels as illustrated in Fig. 4-25. [27] The dashed line indicates that it is a virtual level. Shown in Fig. 4-25 is the interaction of a pump wave with the first excited state of the vibrational mode. The pump wave could also interact with higher-order excited states of the vibrational mode and lead to a larger amount of frequency shift.

As illustrated in Fig. 4-25, Raman scattering is like the situation of three-level atomic systems pumped by an incident light. As shown by the three energy levels, emission of a Stokes wave is more likely to happen than the emission of an anti-Stokes wave. Also, if the pump intensity is not large, only spontaneous emission of the Stokes wave happens. When the pump intensity is very large, there might be stimulated emission of the Stokes wave, called stimulated Raman scattering (SRS), discovered in 1962. [28, 29]

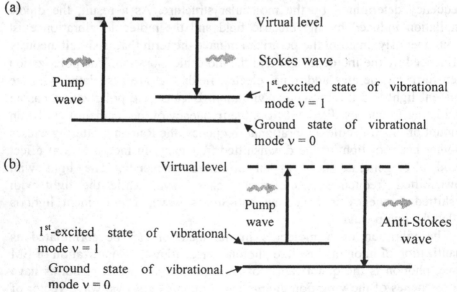

Figure 4-25. Raman scattering represented by the energy levels: (a) generation of Stokes wave; (b) generation of anti-Stokes wave.

With stimulated Raman scattering, the Stokes wave grows with the propagating distance like the usual amplification by stimulated emission in lasers. The growth of the Stokes intensity can be described by the following equation [2, 27]

$$\frac{dI_s}{dz} = g_R I_p I_s \tag{4-9}$$

where I_s is the Stokes intensity, I_p is the pump intensity, and g_R is the Raman gain coefficient. As we described before, Raman scattering is caused by polarization simultaneously influenced by the pump light and the molecule vibration. This effect then contributes to a third-order nonlinear term for the electric susceptibility χ_e. The imaginary part of the third-order nonlinear susceptibility is related to the Raman gain coefficient g_R. With the loss in the fiber taken into account, Eq. (4-9) becomes

$$\frac{dI_s}{dz} = g_R I_p I_s - \alpha_s I_s \tag{4-10}$$

where α_s is the absorption coefficient at the Stokes frequency.

The generation of the Stokes wave is due to the transfer of the energy in the pump wave to the Stokes wave, so the intensity of pump wave varies according to the following equation.

$$\frac{dI_p}{dz} = -\frac{\omega_p}{\omega_s} g_R I_p I_s - \alpha_p I_p \tag{4-11}$$

where α_p is the absorption coefficient at the pump frequency. ω_p and ω_s are the frequencies of the pump wave and the Stokes wave, respectively.

5.3   Light Amplification by Stimulated Raman Scattering

Eq. (4-10) and Eq. (4-11) can be easily solved if the depletion of the pump wave is ignored, i.e. neglecting the first term in the right hand side of Eq. (4-11). For a distance of L, the pump wave and the Stokes wave at the output have their intensities as follows.

$$I_p(L) = I_0 \exp(-\alpha_p L) \tag{4-12}$$

$$I_s(L) = I_s(0)\exp(g_R I_0 L_{eff} - \alpha_s L) \tag{4-13}$$

where I_o is the initial intensity of the pump wave and L_{eff} is given by

$$L_{eff} = \frac{1}{\alpha_p}[1 - \exp(-\alpha_p L)] \tag{4-14}$$

L_{eff} is called Raman effective length within which the pump wave provides a significant contribution to the stimulated Raman emission. Beyond this length, the pump wave is significantly attenuated, so its contribution is reduced. If the fiber loss at the pump wavelength is 0.25 dB/km, the Raman effective length is ~ 20 km.

In optical communications, we are more concerned with the optical power than the light intensity. By integrating Eq. (4-12) and Eq. (4-13) over the cross section of the optical fiber, we obtain the power of the pump wave and the Stokes wave at the output given by

$$P_p(L) = P_0 \exp(-\alpha_p L) \tag{4-15}$$

$$P_s(L) = P_s(0)\exp(g_R P_0 L_{eff} / A_{eff} - \alpha_s L) \tag{4-16}$$

where the power

$$P = \int_0^{2\pi} \int_0^{\infty} I(r,\theta) r \, dr \, d\theta = I \ A_{eff} \tag{4-17}$$

A_{eff} is the effective cross section of the guiding mode in the optical fiber. Here we assume that the pump wave and the Stokes wave have the same field distribution of guiding mode in the optical fiber.

The gain of a Raman amplifier with a length of L is thus given by

$$G_A = \frac{P_s(L)}{P_s(0)\exp(-\alpha_s L)} = \exp(g_R P_0 L_{eff} / A_{eff}) \tag{4-18}$$

If the loss of the pump wave in the optical fiber is very small, the effective length is approximately equal to the actual length of the fiber, $L_{eff} \approx L$. Using typical values Raman gain coefficient $g_R = 1 \times 10^{-13}$ m/W, length of Raman amplifier $L = 1$ km, pump power of $P_o = 500$ mW, and $A_{eff} = 10$ μm^2 in Eq. (4-18), we obtain the Raman gain of more than 20 dB.

If the output power of the Stokes wave is large, then the depletion of pump power is not negligible. Eq. (4-10) and Eq. (4-11) have to be resolved to give correct results. Assuming the fiber loss of the pump wave and the Stokes wave to be the same, we have the saturated gain G_s related to the unsaturated gain G_A by the following equation.

$$G_s = \frac{1+r_0}{r_0 + G_A^{-(1+r_0)}}$$

(4-19)

where r_0 is given by $r_0 = \frac{\omega_p}{\omega_s}\frac{P_s(0)}{P_0} = \frac{n_s^0}{n_p^0}$, the ratio of signal (Stokes) photons to pump photons.

5.4 Pumping for Raman Amplifiers

The configuration of pumping in the Raman amplifier is similar to that in the EDFA. Fig. 4-26 shows a schematic of the Raman amplifier with the pumping source. The fiber has no doping. The spectral range of the gain in the Raman amplifier is determined by the pumping wavelength. Changing the pumping wavelength also changes the spectral range. Thus unlike the EDFA which has a fixed gain spectrum, the Raman amplifier has a flexible choice of gain spectrum. If multiple pumping sources of different wavelengths are used simultaneously, the total gain spectrum becomes very broad. The pumping configuration with multiple pumping sources is illustrated in Fig. 4-27. The light from the four pumping sources at different wavelengths is combined together using a multiplexer and simultaneously delivered to the fiber. The pump wave can also be arranged to propagate along the direction opposite to the signal (Stokes) propagation direction, which is called counter-propagating pumping. The configuration shown in Fig. 4-26 is called co- propagating pumping.

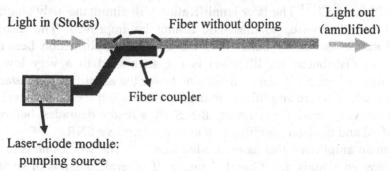

Light in (Stokes)   Fiber without doping   Light out (amplified)

Fiber coupler

Laser-diode module: pumping source

Figure 4-26. Schematic of the Raman amplifier with the pumping source.

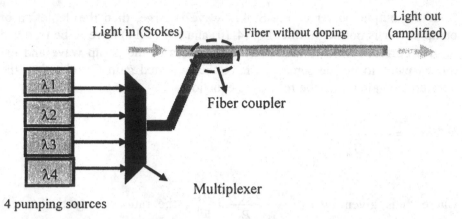

Figure 4-27. Schematic of the Raman amplifier with 4 pumping sources.

5.5   Characteristics of Raman Amplifiers

Because Raman scattering is caused by the vibration of molecule in the medium, the Raman gain and the downshifted frequency depend on the medium property. For a usual fiber, the Stokes wave has a frequency 13.2 THZ lower than the pump wavelength. Because the fiber does not have crystalline structure, its vibrational modes have very broad spectrum. Thus the Raman gain has a broad bandwidth around 60 nm. The Raman gain coefficient of a typical silica-core fiber is about 1×10^{-13} m/W. If the fiber has a high concentration of germanium, the Raman gain coefficient could increase more than five times. The Raman gain is also enhanced if the pump wave is polarized and propagates in a polarization-maintained fiber. The gain could be increased twofold with the polarized pump wave. [30]

There are two advantages of Raman amplifiers over Er-doped fiber amplifiers. The first advantage is the flexibility of choosing the spectral range for amplification. The second is that the gain is distributed over a long distance of fiber, so its signal-to-noise ratio (SNR) is improved. If the power level of signal is already very low, i.e., comparable to the noise level, then the SNR is small. [31] The later amplification will simultaneously amplify the signal and the noise, so the SNR cannot be improved. The distributed amplification is usually better than the discrete amplification because the power in distributed amplification is not attenuated to a very low level. Therefore, a high SNR can be maintained over the amplification distance. In comparison, discrete amplifiers are usually used when the power level of the signal is very low. As a result, the SNR already degrades before it is amplified and the later amplification does not improve SNR.

Raman amplifiers also have disadvantages. [31] The worst one is the cross talk between signals of different channels. If several channels of signals are simultaneously launched into the Raman amplifier, the Raman scattering will cause the conversion of the signal power at the short wavelength to the signal power at the long wavelength. The process is similar to the conversion of the pump power to the signal power. In addition, if a system is used for many channels with closely spaced spectral separation, the above effect will cause the gain of each channel to be different. The gain difference among those channels depends on the channel power, wavelength separation, and the count of the channels.

NOTES

1. Since the gain of the amplifiers, in particular for Er-doped fiber amplifiers and semiconductor optical amplifiers, is the same as the gain of lasers, readers are referred to Chapter 2 for the detail of gain characteristics.
2. For detailed derivation of stimulated Raman scattering, readers are referred to Ref. 1 and Ref. 27.

REFERENCES

1. Agrawal, Govind P. and Dutta, Niloy K., *Semiconductor Lasers*. 2nd Ed., Van Nostrand Reinhold, 1993.
2. Agrawal G. P., *Nonlinear Fiber Optics*. 2/ed. Academic Press, 1995.
3. Miniscalco, W. J., Erbium-doped glasses for fibre amplifiers at 1500 nm. J. Lightwave Technology 1991; 9: 234-250.
4. Armitage, J. R., "Introduction to Glass Fibre Lases and Amplifires." In *Optical Fibre Lasers and Amplifiers*. France, P. W., ed. Blakie and Son Ltd., 1991.
5. Miniscalco, W. J., "Optical and Electronic Properties of Rare Earth Ions in Glass." In *Rare Earth Doped Fiber Lasers and Amplifiers*. Digonnet, M. J. F., ed. Marcel Dekker, Inc., 1993.
6. Siegman, A. E., *Lasers*. University Science Books, 1986.
7. Schmidt, B. E., Mohrdiek, S., and Harder, C. S., "Pump Laser Diodes." In *Optical Fiber Telecommunications IVA: Components*. Kaminow, I. and Li, T., eds. Academic Press, 2002.
8. Sun, Y., Judkins, Srivastava, A. K., Garrett, L., Zyskind, J. L., Sulhoff, J. W., Wolf, C., Derosier, R. M., Gnauck, A. H., Tkach, R. W., Zhou, J., Espindola, R. P., Vengsarkar, A. M., and Chraplyvy, A. R., Transmission of 32-WDM 10-Gb/s channels over 640 km using broad band, gain-flattened erbium-doped silica fiber amplifiers. IEEE Photonincs Technology Letters 1997; 9: 1652-1654.
9. Desurvire, E., "Fundamentals of Noise in Optical Fiber Amplifiers." In Erbium-Doped Fiber Amplifiers, Principles and Applications. Desurvire, E., ed. John Wiley & Sons, Inc., 1994.
10. Mynbaev, D. K. and Scheiner, L. L., *Fiber-Optic Communication Technology*. Prentice Hall, 2001.
11. Ohishi, Y., Kanamori, T., Kitagawa, T., Takahashi, S., Snitzer, E., and Sigel, G. H., Pr^{3+}-doped fluoride fiber amplifier operating at 1.31 μm. Optics Letters 1991; 16: 1747-1749.

12. Nishida, Y., Tamada, M., Kanamori, T., Kobayashi, K., Temmyo, J., Sudo, S., and Ohishi, Y., Development of an efficient praseodymium-doped fiber amplifier. IEEE Journal of Quantum Electronics 1998; 8: 1332-1339.

13. Olsson, N. A., Oberg, M. G., Tzeng, L. D., and Cella, T., Ultra-low reflectivity 1.5 μm semiconductor laser amplifiers. Electronics Letters 1988; 24: 569-570.

14. Alphonse, G. A., Gilbert, D. B., Harvey, M. G., and Ettenberg, M., High-power superluminescent diodes, IEEE Journal of Quantum Electronics 1988; 24: 2454-2457.

15. Lin, C. F., The influence of facet roughness on the reflectivities of etched-angled facets for superluminescent diodes and optical amplifiers. IEEE Photonics Technology Letters 1992; 4: 127-129.

16. Semenov, A. T., Shidlovski, V. R., and Safin, S. A., Wide spectrum single quantum well superluminescent diodes at 0.8 mm with bent optical waveguide. Electronics Letters 1993; 29: 854-856.

17. Lin, C. F. and Juang, C. S., Superluminescent diode with bent waveguide. IEEE Photonics Technology Letters 1996; 8: 296-208.

18. Lin, C. F., Superluminescent diodes with angled facet etched by chemically assisted ion beam etching. Electronics Letters 1991; 27: 968-969.

19. Cha, I., Kitamura, M., Honmou, H., and Mito, I., 1.5 μm band traveling-wave semiconductor optical amplifiers with window facet structure. Electronics Letters 1989; 25: 1241-1242.

20. Mukai, T. and Yamamoto, Gain, frequency bandwidth, and saturation output power of AlGaAs DH laser amplifiers. IEEE Journal of Quantum Electronics 1981; 17: 1028-1034.

21. Feuer, M. D., Wiesenfeld, J. M., Perino, J. S., Burrus, C. A., Raybon, G., Shunk, S. C., and Dutta, N. K., Single part laser amplifier modulators for local access. IEEE Photonics Technology Letters 1996; 8: 1175-1177.

22. Tai, C. and Way, W. I., Dynamic range and switching speed limitation of an N x N optical packet switch based on low gain semiconductor optical amplifiers. IEEE Journal of Lightwave technology 1996; 14: 525 -533.

23. Jourdan, A., Masetti, F., Garnot, M., Soulage, G., and Sotom, M., Design and implementation for a fully reconfigurable all-optical crossconnect for high capacity multi wavelength transport networks. IEEE Journal of Lightwave technology 1996; 14:1198-1206.

24. Raman, C. V., A new radiation. Indian Journal of physics 1928; 2: 387-398.

25. Kittel, Charles, *Introduction to Solid State Physics*. John Wiley & Sons, Inc., 1976.

26. Tang, C. L., Saturation and spectral characteristics of the Stokes emission in the stimulated Brillouin process. Journal of Applied Physics 1966; 37: 2945-2955.

27. Yariv, A., Quantum Electronics. 2nd Ed., John Wiley & Sons, Inc., 1977.

28. Woodbury, E. J. and Ng, W. K., Ruby laser operation in the near IR. Proc. IRE 1962; 50: 2367.

29. Shen, Y. R. and Bloembergen, N., Theory of stimulated Brillouin and Raman scattering. Physical Review 1965; 137: A 1787-A1804.

30. Stolen, R. H. Polarization effects in fiber Raman and Brillouin lasers. IEEE Journal of Quantum Electronics 1979; 15: 1157-1161.

31. Rottwitt, K. and Ztentz, A., "Raman Amplification in Lightwave Communication Systems." In *Optical Fiber Telecommunications IVA: Components.* Kaminow, I. and Li, T., eds. Academic Press, 2002.

Chapter 5

PHOTODETECTORS AND DETECTION OF OPTICAL RADIATION

1. INTRODUCTION

When optical signals transmitted in optical fiber arrive at the destination, they have to be converted to electronic signals. Photodetectors are used for this purpose.[1, 2] The converted electronic signals are then further amplified by electronic amplifiers or demodulated to reproduce the original signals from the transmitter side. If the optical signals arriving at the photodetectors are very weak, they may be amplified by an optical preamplifier before they are sent to the photodetectors. In this chapter, we will first focus on the working principle of photodetectors that convert the optical signals to electronic signals. [3] Several types of photodetectors commonly used for optical communication will be introduced. Then the sources of noise associated with signal detection will be also explained. Finally, an evaluation of the signal-to-noise ratio will be discussed.

1.1 Quantum detection

In Chapter 1, we explained that light exhibits behaviors similar to a particle. Light detection can be better explained from the particle nature of light. From this point of view, the photodetector, working as a quantum detector, receives photos from its input and generate electrical carriers (electrons or holes) at its output, as shown in Fig. 5-1. Therefore, there is a direct correspondence between the number of photons absorbed by the photodetector and the number of carriers generated.[3] Ideally, receiving one photon, the quantum detector generates one carrier. In practice, there usually more photons absorbed than carriers generated. The ratio of the carriers generated per second to the photons absorbed per second is defined as the quantum efficiency.

$$\eta_{qe} = \frac{number \ of \ carriers \ generated \ per \ second}{number \ of \ photons \ absorbed \ per \ second}$$

$$= \frac{i/e}{P_{abs}/h\nu} \qquad\qquad (5\text{-}1)$$

where i is the current at the output of the detector; P_{abs} is the optical power received by the detector; e is the charge of an electron and $h\nu$ is the energy of a single photon incident on the detector.

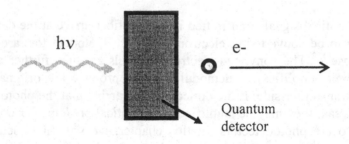

Figure 5-1. A photodetector works as a quantum detector.

2.   PHOTODETECTORS

There are many types of photodetectors, including the vacuum photodiode, photomultiplier (PMT), photoconductor, PN-photodiode, PIN-photodiode, and avalanche photodiode (APD). For optical communications applications, the PIN-photodiode and the APD are the two most commonly used because they have good quantum efficiency and are made of semiconductors with potentially low-cost in mass production. On the other hand, the PN-photodiode is the simplest structure that has the basic physics of photo detection. It is better to understand the principle of PN photodiodes before getting into the detail of PIN-photodiodes and APD. So we will start with PN photodiodes before taking on PIN-photodiodes and APD photodiodes.

2.1   PN-Photodiode

PN-photodiodes are made of semiconductors.[4] As we explained in Chapter 2, when the n-type semiconductor and the p-type semiconductor are connected together, electrons in the n-type semiconductor and holes in the p-type semiconductor will diffuse to the opposite side. A depletion region that has very few electrons and holes is formed. In the depletion region, there are also many immobile ions called space charge. This establishes an electric field as shown in Fig. 5-2.[5] In equilibrium, the electric field is just strong enough to prevent further diffusion of electrons and holes to the opposite side. That is, the electric field forces electrons to move toward the n-type semiconductor. Such movement is in balance with the diffusion of electrons toward the p-side. Similarly, holes have the balance of movement due to electric field and diffusion. Consequently, there is no net flow of electrons and holes across the p-n junction as the equilibrium is reached. The electric field in the depletion region then causes a voltage drop between the p-type semiconductor and the n-type semiconductor. This voltage is called built-in voltage (V_{bi}). eV_{bi} is slightly less than the bandgap energy of the semiconductor. On the other hand, the n-type and p-type semiconductors outside the depletion region have no immobile ions, so there is no electric field and no voltage drop there when no external bias is applied.

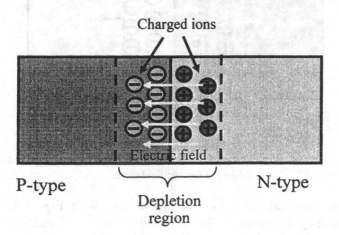

Figure 5-2. The charged ions and electric field in the depletion region of the PN-photodiode.

When light is incident on the PN-photodiode, the semiconductor absorbs the photons and electron-hole pairs are generated. What will happen to these extra electron-hole pairs generated by light incidence? The extra electron-hole pairs will behave differently, depending on whether they are generated in the depletion region or outside the depletion region.

If the extra electrons and holes are generated in the n-type semiconductor, because there is no electric field outside the deletion region, the electrons and holes are not separated. Instead, the holes will soon recombine with electrons within the minority carrier lifetime, which is in the order of 1 ns. Similarly, if the extra electrons and holes are generated in the p-type semiconductor, the electrons will soon recombine with holes. In other words, the extra electrons and holes will soon disappear if they are generated outside the deletion region. Those generated electrons and holes have no contribution to photo current, defined as current generated by light incidence.

On the other hand, if extra electrons and holes are generated in the depletion region, the electric field there will force electrons to move toward the n-type semiconductor. Holes are also forced to move toward the p-type semiconductor. Therefore, the generated electron and hole are separated. The situation is illustrated in Fig. 5-3. As a result, electrons in the n-type semiconductor and holes in the p-type semiconductor exceed the equilibrium value.

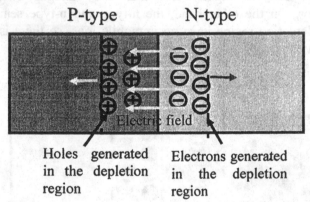

P-type   N-type

Holes generated in the depletion region   Electrons generated in the depletion region

Figure 5-3. Electrons and holes generated in the depletion region are forced to separate by the electric field there.

When an external circuit is set up, as shown in Fig. 5-4, the extra electrons will pass through the external circuit and reach the p-type semiconductor. Those electrons then recombine with the extra holes in the p-type semiconductor. The flow of electrons thus leads to current in the external circuit.

2.1.1   Operation of PN-Photodiode in Open Circuit

If there is no external circuit or the external circuit is left open, the extra electrons will not be able to move to the p-type semiconductor. Then the

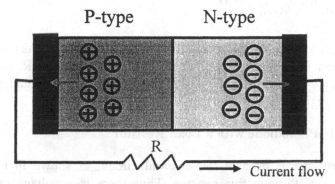

Figure 5-4. Electrons move to p-type semiconductor through the external circuit, leading to current flow.

generated electrons and holes accumulate in the n-type semiconductor and the p-type semiconductor, respectively. The accumulated electrons and holes again establish an electric field, which has the opposite direction to the electric field built in the depletion region. That is, another voltage is built between the n-type semiconductor and the p-type semiconductor. This voltage is equivalent to a forward bias of the p-n junction. If the intensity of the incident light is large, the generated electrons and holes could establish an electric field comparable to the electric field built in the depletion region. The two fields cancel one another. Then further generation of electrons and holes due to the incidence of light cannot be forced to separate by the electric field in the depletion region. In this case, there is no current flow in the depletion region, n-type semiconductor, and p-type semiconductor. The voltage V built between the n-type semiconductor and the p-type semiconductor can be calculated according to the following relation.

$$I = I_0[\exp(\frac{eV}{kT}) - 1] - I_{pc} = 0 \qquad (5\text{-}2)$$

where $I = I_0[\exp(\frac{eV}{kT}) - 1]$ is the relation of the current of a p-n diode under a bias voltage V; I_0 is the reverse-bias current, which is very small; I_{pc} is the photocurrent. According to Eq.(5-1), the photocurrent is given by

$$I_{pc} = \eta_{qe} \bullet e \bullet \frac{P_{abs}}{h\nu} \qquad (5\text{-}3)$$

Thus the open-circuit voltage of the PN-photodiode is given by

$$V_{oc} = \frac{kT}{e} \ln(-\frac{e\eta_{qe}P_{abs}}{h\nu I_0} + 1)$$ (5-4)

The PN-photodiode can be operated in open-circuit to work as a solar cell. Such operation is called photovoltaic mode.[6] The photovoltaic voltage is given by Eq. (5-4). This voltage is no more than the built-in voltage, V_{bi}, in the depletion region.

2.1.2   PN-Photodiode with a Load Resistor

When a resistor is used in the external circuit, as shown in Fig. 5-4, current will pass through this resistor. Thus there is a voltage drop, V_0, across the resistor.

$$V_0 = IR$$ (5-5)

where I is the current in the external circuit. It is equal to the photocurrent minus the diode current.

$$I = I_{pc} - I_0[\exp(\frac{eV_0}{kT}) - 1]$$ (5-6)

From Eq. (5-3), Eq. (5-5), and Eq. (5-6), we are able to solve the current I and voltage drop V_0 in terms of the power of the incident light although the mathematics are not easy.

On the other hand, if the voltage drop across the resistor is small, say eV_0 not much larger than kT, then the diode current $I_0[\exp(\frac{eV}{kT}) - 1]$ is small compared to the photocurrent I_{pc}. The diode current is negligible, so $I \sim I_{pc}$. In this case, the measured voltage across the resistor is proportional to the photocurrent, which is proportional to the power of the incident light, P_{abs}.

$$V_0 = I_{pc}R$$ (5-7)

2.1.3 PN-Photodiode under Reverse Bias

If the photocurrent is large due to an intense incident light, the voltage V_0 in Eq. (5-7) becomes large. Similar to the situation of the open circuit, this voltage performs like a forward bias for the p-n diode. When it is comparable to the built-in voltage, the electric field established by this voltage is also comparable to the electric field built in the depletion region. Then according to Eq. (5-6), the current in the external circuit is much smaller than the photocurrent. The measured current or the measured voltage in the external circuit is no longer proportional to the power of the incident light. Therefore the set up shown in Fig. 5-4 is valid only for the incident light of low power. If the optical power is large, a reverse bias is usually added in the external circuit to avoid the above situation. The circuit is shown in Fig. 5-5.

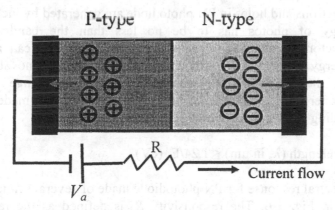

Figure 5-5. The PN-photodiode under reverse bias.

With the reverse-bias voltage V_a, the diode current becomes $I_0[\exp(\frac{IR-eV_a}{kT})-1]$. The current in the external circuit is also equal to the photocurrent minus the diode current.

$$I = I_{pc} - I_0[\exp(\frac{IR-eV_a}{kT})-1] \tag{5-8}$$

As long as $IR < e\,V_a$, the diode current (the second term of the right hand side in Eq. (5-8) is negligible, so the external current equals the photocurrent, $I = I_{pc}$. As a result, the measured current or the measured voltage is proportional to the power of the incident light again. Using the relation $I_{pc}\,R$

$< e\ V_a$, where $I_{pc} = \eta_{qe} \bullet e \bullet \dfrac{P_{abs}}{h\nu}$, we are able to calculate the limit of the optical power that can be measured with the set up shown in Fig. 5-5.

$$P_{abs} < \frac{h\nu V_a}{\eta_{qe} R} \tag{5-9}$$

The reverse bias has other effects as well. As we explained in Chapter 2, the depletion region increases with reverse bias. Because only carriers generated in the depletion region contribute to photocurrent, the reverse bias thus increases the region of PN-photodiode to interact with the incident light.

2.1.4   Spectral Response of PN-Photodiodes

The electrons and holes in PN-photodiode are generated by incident light. The energy of photos has to be no less than the bandgap of the semiconductor, so bonded electrons in the valence band can absorb the photon energy to transit to the conduction band and become movable carriers. Therefore, the PN-photodiode works for the spectral range above the bandgap energy (E_g) of the semiconductor that it is made of. The corresponding wavelength follows the following relation.

$$\text{wavelength } (\lambda, \text{ in } \mu m) \le 1.24/E_g \text{ (eV)} \tag{5-10}$$

The spectral response for PN-photodiode made of several semiconductors is shown in Fig. 5-6. The responsivity R_0 is defined as the ratio of the generated photocurrent to the optical power absorbed by the photodetector.

$$R_0 = \frac{I_{pc}}{P_{abs}} = \frac{e\eta_{qe}}{h\nu} = \frac{\eta_{qe}}{1.24} \lambda \ (\frac{A}{W}) \tag{5-11}$$

where λ is in the unit of μm. [2, 7]

Si has a response from 0.3 μm to 1.1 μm. Ge has a response from 0.5 μm to 1.8 μm. The response for InGaAs is from 1.0 μm to 1.8 μm. The semiconductor InGaAs is the most popular material to make photodetectors for fiber optic communications because its spectral response is mainly in the same spectral region.

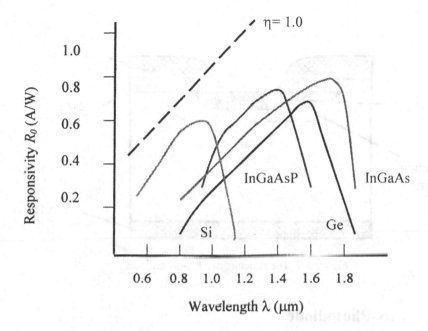

Figure 5-6. Spectral response of photodetector made of several materials.

2.1.5 Geometrical Structure of PN-Photodiode

The structure of a PN-photodiode is schematically shown in Fig. 5-7. The fabrication is very simple. The n-type wafer is ion-implanted with another dopants to become p-type semiconductor. The top p+ layer is made very thin. Light is incident on detector from the p-side, passing through the p+ layer, and absorbed in the depletion region. In fact, the depletion region is very thin, which is about 20 nm, so only very small amount of light is absorbed in this region. Most of light is absorbed either in the p-type semiconductor or in the bottom n-type semiconductor. Therefore, the PN-photodiode does not have good responsivity. To improve the situation, the PIN structure is used now for applications in the real world. The PIN-photodetector will be described in the following section.

Fig. 5-7 shows that the light is incident from the top. It is also possible that light is incident from the direction parallel to the p-n junction. However, due to the thin depletion region, the portion of light absorbed there is still very small.

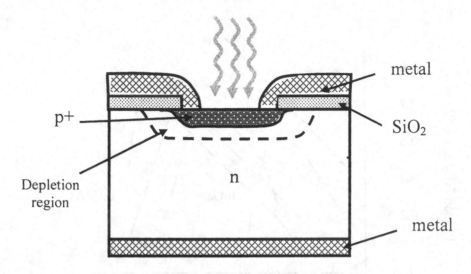

Figure 5-7. The structure of a PN-photodiode.

2.2   PIN-Photodiode

As we discussed in the last section, only those carriers that are generated in the depletion have contribution to the photocurrent. For a PN-photodiode, the depletion region is too thin to have significant light absorbed there. As a result, the quantum efficiency is low. PIN-photodiodes improve the quantum efficiency. An intrinsic layer, indicated as i, is placed between the p-type semiconductor and the n-type semiconductor. Fig. 5-8 shows the band structure of the PIN-photodiode without external bias. There is only slight band bending at the interface of the p-i junction and the i-n junction. When the PIN-photodiode is under reverse bias, the entire intrinsic layer could become the depletion region. Its band structure is shown in Fig. 5-9. The band bending extends all the way through the i-layer. The electric field is also plotted there to show that the entire region has the strong field to force electrons and holes generated in this region to separate and move toward the n-layer and the p-layer, respectively.

2.2.1   Thickness of the Intrinsic Layer

The i-layer is thick, so the PIN-photodiode has a very thick depletion region when it is under reverse bias. This layer can absorb more than 80 % of the incident optical power if the i-layer is thick enough. The requirement of the thickness of the i-layer can be estimated as follows. As shown in

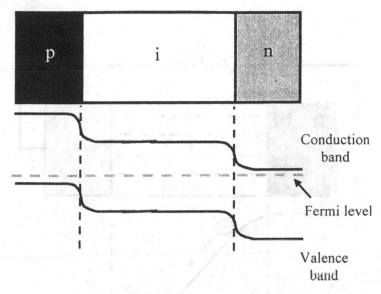

Figure 5-8. Band structure of the PIN-photodiode across the junctions.

Fig. 5-10, the light power in the material exponentially decays with the propagation distance. At the distance of w_p, the light enters the i-layer. Its power is $P_{in} \exp(-\alpha w_p)$, where α is the absorption coefficient of the semiconductor. As it leaves the i-layer, the power decays to $P_{in} \exp[-\alpha(w_p+w_i)]$. Therefore, the amount absorbed in the i-layer is $\{P_{in} \exp(-\alpha w_p) - P_{in} \exp[-\alpha(w_p+w_i)]\}$. The ratio of light absorbed in the i-layer is

$$\text{Absorption Ratio} = \{P_{in} \exp(-\alpha w_p) - P_{in} \exp[-\alpha(w_p+w_i)]\}/ P_{in}$$

$$= \exp(-\alpha w_p) - \exp[-\alpha(w_p+w_i)] \tag{5-12}$$

For $w_p \ll \alpha^{-1}$, if $w_i = 2\ \alpha^{-1}$, more than 80 % of optical power is absorbed in the i-layer, resulting in a very good quantum efficiency. For $In_{0.53}Ga_{0.47}As$, the absorption coefficient at 1.3 μm and 1.55 μm is about 10^4 cm^{-1} and 0.7×10^4 cm^{-1}, respectively. To achieve the above condition $w_i = 2\ \alpha^{-1}$, the thickness of this region is only about 2-3 μm. For Si, the absorption coefficient in the spectral range of 0.8 μm - 0.9 μm is less than 10^3 cm^{-1}. To achieve the above condition $w_i = 2\ \alpha^{-1}$, the thickness of this region has to be between 20 μm and 50 μm.

(a)

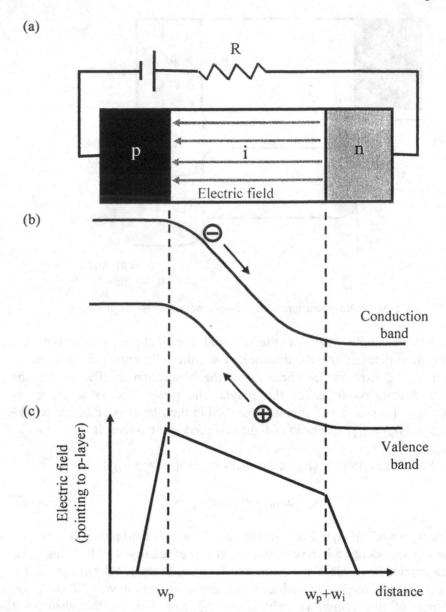

(b)

(c)

Figure 5-9. (a) Circuit of PIN-photodiode under reverse bias. (b) Band structure of the PIN-photodiode across the junctions under reverse bias. (c) Electric field.

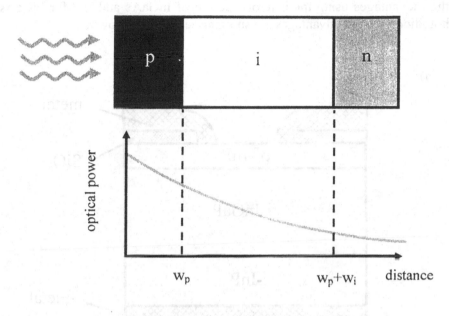

Figure 5-10. Variation of light power along the penetration distance.

2.2.2   InGaAs PIN-photodiode

When the intrinsic layer (i-layer) is made of InGaAs semiconductor, it can be sandwiched between n-type and p-type InP semiconductors. Such a structure can be easily formed using normal epitaxial growth technology like MBE or MOCVD because $In_{0.53}Ga_{0.47}As$ is lattice matched to InP. The $In_{0.53}Ga_{0.47}As$ has a bandgap of 0.74 eV, which corresponds to wavelength of 1.68 μm, so this intrinsic layer has spectral response for the entire bandwidth of optical communication. The InP semiconductor for n-layer and p-layer has a bandgap of 1.35 eV, corresponding to wavelength of 0.92 μm. Therefore, no light is absorbed by the n-layer or the p-layer, increasing the power absorbed by the i-layer. Then light may be incident from the top p-layer or from the bottom n-layer, regardless of their thickness. These situations are illustrated in Fig.5-11.

In Fig.5-11(a), light is incident on the top layer. It has a structure similar to the PIN-photodiode made of Si. The advantage is that there is no absorption of light in p-layer, so the quantum efficiency can be increased. In Fig. 5-11(b), light is incident from the bottom n-layer. Because InP is transparent to the spectral range for optical communication, there is no light absorption in this layer either, even though it is very thick. There are also other advantages using the hetero-structure of InGaAs and InP for InGaAs photodiodes. Their advantages are summarized in the following.

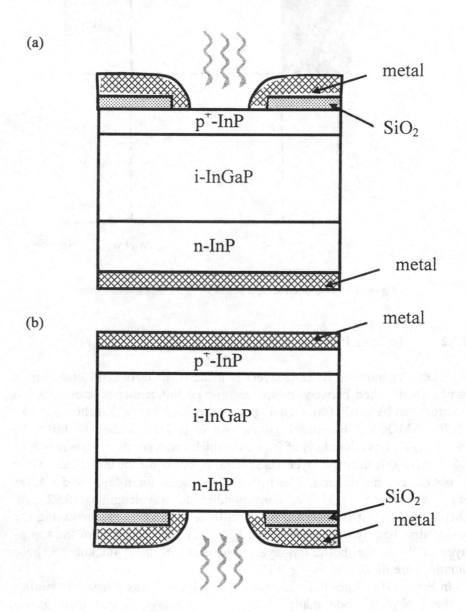

Figure 5-11. The structure of InGaAs PIN-photodiode for light incidence from (a) the top p-layer and (b) the bottom n-layer.

- The InP semiconductor is transparent, so there is no light absorption in the n-layer or p-layer, increasing light absorbed in the i-layer.
- Because n-layer and p-layer do not absorb light, no carriers are generated there. Thus there is no diffusion of carriers into the depletion region. This could increase the response time.
- Forming a hetero-structure with a wide bandgap semiconductor, the breakdown voltage is increased and the reverse bias current is reduced at the same time. For p-n junctions made of narrow bandgap semiconductors, the breakdown voltage is low, leading to easy damage of the diode. In addition, the reverse bias current (I_0) is also large because I_0 is proportional to exp (-Eg/kT).
- The InGaAs layer is epitaxially grown on the InP substrate, which can be semi-insulating. The semi-insulating substrate allows for the fabrication of bonding pads with low-capacitance, so the response frequency can be increased.
- Because light can be reflected by the top metal layer, it can pass through the i-layer twice. This is illustrated in Fig.5-11(b), This means that the thickness of the i-layer can be reduced to maintain the same responsivity. Because the transit time of carriers across the depletion region is reduced, the response speed is increased.

2.2.3 Edge Illumination Structure of PIN-photodiode

An influential factor for the response time of the PIN-photodiode is the transit time of electrons and holes across the depletion region. To increase the response speed, i.e., to reduce the response time, the depletion region should be narrow. However, that will then reduce the responsivity, even if we use light incident from the top p-layer or from the bottom n-layer, as shown in Fig. 5-11. To avoid the limitation caused by the tradeoff between the response time and the responsivity, another structure is proposed. Fig. 5-12 shows the new structure with edge illumination.[8, 9] The i-layer is still sandwiched between the p-type semiconductor and the n-type semiconductor. For InGaAs PIN-photodiode, it is particularly advantageous to use InGaAs for the i-layer and InP for the n-layer and the p-layer. Because InGaAs has a larger refractive index than the InP, the structure becomes a waveguide for light guiding in the InGaAs layer when light is incident from the edge.

The edge-illumination structure has other advantages. First, it has a similar waveguide structure to the edge-emission laser diodes that are commonly used for optical-fiber communication nowadays. A fiber-pigtailed package similar to the fiber-pigtailed LDs and SOAs can be used for this type of PIN-photodiode. Second, the electrode can be designed in such a way that the electronic signal on the electrode can be velocity-matched to the optical signal propagating in the waveguide. Thus the operation speed can be very fast, leading to a very large bandwidth. It is not uncommon to achieve 110GHz.[10]

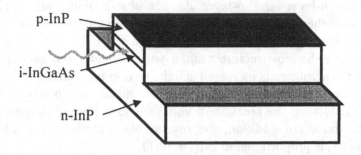

p-InP

i-InGaAs

n-InP

Figure 5-12. PIN-photodiode for edge illumination.

2.2.4   Response Time

The response speed of a PIN-photodiode is limited mainly by two factors. First is the transit time of carriers across the depletion region. The second is the RC time constant associated with the electrical parameters of the electrical circuit containing the photodiode and the loading resistance.[3]

Because only the carriers generated in the depletion region contribute to the photocurrent, the time for electrons or holes to transport from one side of the depletion region to the other side sets up a limit of the response speed of the device. If electrons generated by the first optical pulse are still in the depletion region while the second pulse has arrived at the same region, the electrons generated by the second pulse will mix with those generated by the first pulse. This leads to an ambiguity in distinguishing which electronic signals correspond to each optical pulse. To avoid the ambiguity of the electronic signals, the two optical pulses should temporally separate at least for the transit time across the depletion region.

The carriers in the depletion region are pulled by the electric field present and drift toward the n-layer or the p-layer. When the electric field is very large, the velocities of carriers in semiconductors tend to saturate. The electric field in the depletion region is usually large enough to have the carries reach their saturation velocity in most of the drift length, particularly for devices under a reverse bias. Thus the velocity of electrons or holes can be treated as a constant. The time for carriers to transit through the depletion region is then simply given by

$$\tau_{transit} = \frac{w_i}{v_{sat}} \tag{5-13}$$

where v_{sat} is the saturation velocity. It is usually about 5×10^6-10^7 cm/sec, depending on the semiconductor and the type of carriers. For high speed InGaAs PIN-photodiodes, the depletion region may be as short as 0.5 μm. The transit time will then be only a few ps.

To understand the second limiting factor, namely the RC time constant of the PIN-photodiode and its associated circuit, we need to know the electrical parameters of the PIN-photodiode. A PIN-photodiode can be considered to be comprised of a current source (current generated by light incidence) and a combination of capacitance and resistance. Its equivalent circuit is shown in Fig. 5-13. Within the dashed line are the electrical parameters representing the photodiode. C_j is the capacitance of the junction across the p-layer to the n-layer. R_{rev} is the resistance of a reverse-biased diode. R_s is the series resistance for the n-layer and the p-layer.

The junction capacitance is an important parameter that limits the response speed. When the PIN-photodiode is under strong reverse bias, the entire intrinsic layer is depleted. Then the junction capacitance is determined by the depletion region, mainly formed by the intrinsic layer.

$$C_j = \frac{\varepsilon_r \varepsilon_0 A}{w_i} \qquad (5\text{-}14)$$

where ε_r is the relative dielectric constant and A is the junction area.

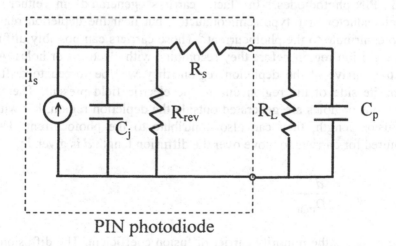

PIN photodiode

Figure 5-13. Equivalent circuit of a PIN-photodiode. The electrical parameters of the PIN-photodiode are in the dashed line.

The external circuit shown in Fig. 5-13 consists of the load resistance R_L and the parasitic capacitance, which is caused by the bonding pads or package. From the circuit shown in Fig. 5-13, we can derive the response of the voltage across the load resistor, V_L, with respect to the photocurrent i_s.

$$\frac{V_L}{i_s} = \frac{R_L}{1 + j\omega(R_s C_j + R_L C_j + R_L C_P) - R_S R_L C_j C_P \omega^2} \qquad (5\text{-}15)$$

The time constant associated with the circuit is given by

$$\tau_{RC} = R_S C_j + R_L C_j + R_L C_p \qquad (5\text{-}16)$$

The total response time is given by

$$\tau^2 = \tau_{transit}^2 + \tau_{RC}^2 \qquad (5\text{-}17)$$

2.2.5   Response due to Carrier Diffusion into the Deletion Region

In the previous sections, we state that only carriers generated in the depletion region contribute to the photocurrent. This simplification was made in order to clearly explain the working principles of PN-photodiodes and PIN-photodiodes. In fact, carriers generated in either n-type semiconductor or p-type semiconductor, but near the depletion region can also contribute to the photocurrent.[3] Those carriers can possibly diffuse into the diffusion region before they recombine with electrons or holes. As soon as they arrive at the depletion region, they will be forced to drift to the opposite side of the region due to the electric field present. Therefore, if electrons or holes are generated outside the depletion region, but within the diffusion length, they can also contribute to the photocurrent. The time required for carriers to move over the diffusion length d is given by [11]

$$\tau_{diff} = \frac{d^2}{2D_{min}} \qquad (5\text{-}18)$$

where D_{min} is the minority carrier diffusion coefficient. The diffusion length is usually in the order of micrometers. The diffusion time then may be in the order of nanoseconds. Those carriers may lead to a long tail on the actual electronic pulse, as shown in Fig. 5-14. Those carriers generated outside the depletion region are very small in number for InGaAs PIN-photodiodes because the p-layer and the n-layer have negligible absorption. Therefore, the photocurrent generated by such carriers causes very few problems.

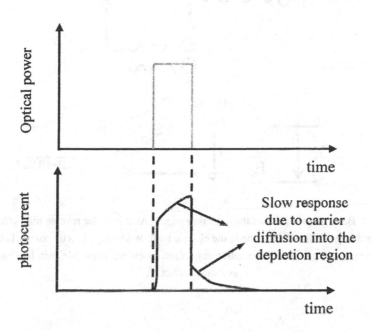

Figure 5-14. (a) A step optical pulse. (b) The corresponding electronic pulse shape. The tail is due to the carriers generated near the depletion region within the diffusion length.

2.3   Avalanche (APD) Photodiode

When a p-n structure is under very strong reverse bias, the variation of the conduction band and the valence band across the depletion region becomes very large. Its band structure is shown in Fig. 5-15. In this situation carriers that transport across the depletion region can obtain very large kinetic energy. Those energetic carriers may potentially collide with the crystal lattice and lose part of their energy to the lattice atoms. If their energy before collision is larger than the bandgap energy of the valence electrons, the collision can free the electron that is bonded to the lattice atom, creating a new electron-hole pair. Such process is called impact ionization. The newly created pair of the electron and hole is forced to separate by the strong electric field again. The new electron and hole can then obtain sufficient kinetic energy due to the electric field to cause another impact ionization. Subsequent impact ionization can happen repeatedly, leading to an avalanche effect.[12] Therefore, many electrons and holes more than the original electron-hole pair generated by photo incidence are created by the avalanche effect. The process is also shown in Fig. 5-15.

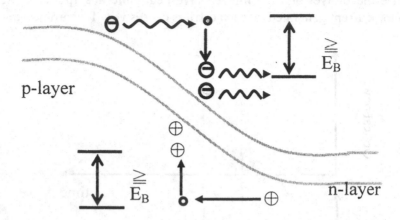

Figure 5-15. Band diagram across the depletion region. As it is under reverse bias, electron and hole gain large drift velocity due to the electric field. When their kinetic energy is larger then the bandgap energy, impact ionization may create a new electron-hole pair, leading to an avalanche effect.

2.3.1   Multiplication Factor of APD

Because the total carriers generated in the APD photodiode are more than those directly generated by light absorption, there is a multiplication M of the APD photodiode.

$$M = \frac{I_M}{I_P} \qquad\qquad (5\text{-}19)$$

where I_M is the multiplied output current and I_P is the unmultiplied photocurrent. In fact, every multiplication from the first pair of electron and hole is not the same, so the multiplication is an average value. The multiplication can be as high as 1000. It is a function of the bias voltage, as shown in Fig. 5-16. For practical applications, APD is operated to have the multiplication of 100. The reverse bias voltage ranges from 10 V to 100V.[7] If the bias voltage is too high, the photodiode may be in danger of creating self-sustaining avalanche current without photoexcitation, leading to extra noise from the photodiode.

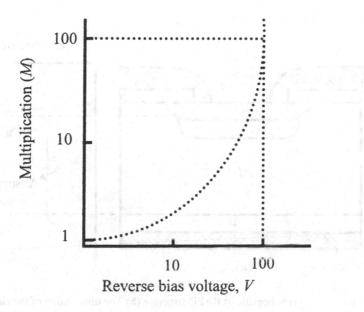

Figure 5-16. Typical variation of multiplication with the reverse bias voltage.

2.3.2   Reach-Through Avalanche Photodiode (APD)

A practical type of APD is the reach-through avalanche photodiode (RAPD). Its schematic of structure is shown in Fig. 5-17(a). The corresponding electric field is also plotted in Fig. 5-17 (b). The device consists of p+-π-p-n layers. The π layer is basically an intrinsic semiconductor, working for light absorption. The electric field extends all the way through the π region to the pn+ junction, so it is called reach through avalanche photodiode. However, the electric field in the π layer is relatively low, compared to the field in the p layer. Electrons and holes generated in the π layer by photoexcitation separate due to this field. Electrons then drift to the pn+ junction that has a very strong electric field for impact ionization. It is advantageous to have only electrons for impact ionization.[13]

For APDs working at wavelengths up to 1.6 μm, the semiconductors used have smaller bandgap than Si. In this case the dark current increases. Although it helps to use cooling techniques to reduce dark current, this inevitably increases the complexity of the photodiode and the operation cost. The situation can be improved by separating the absorption and the gain regions. It is thus called separate absorption and multiplication avalanche photodiode (SAM-APD). The absorption layer is made of $In_{0.53}Ga_{0.47}As$, while the layer for avalanche multiplication is made of InP. A schematic of the device structure is illustrated in Fig. 5-18.

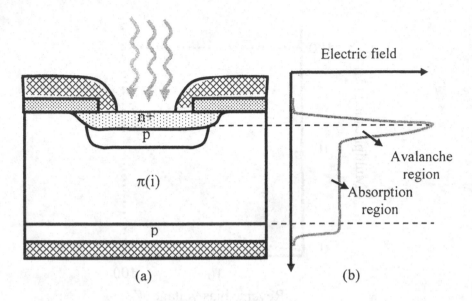

Figure 5-17. (a) Schematic of RAPD structure.(b) The distribution of the electric field.

2.3.3   Discussion on Characteristics of APD

Because the APD has a region of strong electric field for impact ionization, it takes time to go through the avalanche process. The response time is thus degraded by approximately a factor of two.[14] In addition, in order to have the strong electric field for multiplication, the operation voltage is much larger than the usual PIN-photodiode and other electronics. This may cause the difficulty of circuit integration of APD with electronics or increase operation cost because of the inclusion of the high-voltage electronics.

The multiplication of APD also strongly depends on the temperature.[15] In order to maintain the operation temperature, it is necessary to use a thermoelectric cooler. If there is no temperature control, it will be necessary to have an electronic circuit to automatically adjust the bias in order to keep the multiplication constant. In either case, the cost is increased.

In comparing the detection sensitivity of the APD with the PIN-photodiode, because the APD has much larger gain, it offers a better sensitivity. The sensitivity increase may be on the order of 4-7 dB over PIN photoreceivers even though the APD also has additional noise induced during the impact ionization. This advantage is however not better than the PIN-photodiode with an optical preamplifier, which could provide about 12-15 dB sensitivity margin better than the PIN alone. However, using an APD is less costly than using an optical preamplifier in front of the PIN-photodiode.

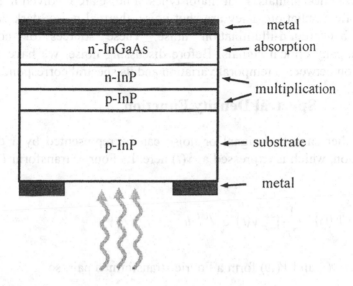

Figure 5-18. Schematic structure of SAM-APD.

On the other hand, a usual PIN plus an electronic amplifier containing field-effect transistor (FET) circuitry could also provide good sensitivity margin. This circuit amplifies the electronic signal before the noise due to the load resistor comes in, so the signal to noise (S/N) ratio is good. However, the response time for the entire circuitry consisting of the PIN-photodiode and the electronic amplifier is much slower than that of the PIN-photodiode or APD alone. For data transmission below one gigabit per second, the combination of a typical PIN and an electronic amplifier may be advantageous over APD for its low cost and because it does not require voltage above 5V.

3. NOISE INVOLVED IN THE DETECTION OF OPTICAL RADIATION [17]

If a photodetector can detect the light without limitation on the minimum optical power, it is not necessary to amplify the optical signal. Unfortunately, there is always minimum signal power that a photodetector can detect. Such minimum signal power is defined as the sensitivity. Then we have a further question. Why is there limitation on the minimum optical power that a photodetector can detect? The reason is noise. If the signal power is less than the noise power, we will not be able to distinguish the signals clearly. Therefore it is important to understand noise and its source when we want to detect optical signals. Four major types of noise are involved in the detection of optical radiation. They are shot noise, thermal noise, dark current noise, and background-illumination noise. These sources of noise always accompany optical signals. Before discussing noise, we have to know the relation between a temporal variation and its spectral correspondence.

3.1   Spectral Density Function

Either an actual signal or noise can be represented by a time varying function, which is expressed as $v(t)$ here. Its Fourier transform $V(\omega)$ is given by

$$V(\omega) = \frac{1}{2\pi} \int_{-\infty}^{\infty} v(t) \, e^{-j\omega t} dt \tag{5-20}$$

In fact, $v(t)$ and $V(\omega)$ form a Fourier-transformed pair, so

$$v(t) = \int_{-\infty}^{\infty} V(\omega) \, e^{j\omega t} d\omega \tag{5-21}$$

In practice, we are not able to measure the temporal variation for infinite time. Thus we consider the function $v(t)$ to be zero when $t \leq -T/2$ and $t \geq T/2$, so the integration is evaluated for the time between $-T/2$ and $T/2$. Eq.(5-20) then becomes

$$V_T(\omega) = \frac{1}{2\pi} \int_{-T/2}^{T/2} v(t) \, e^{-j\omega t} dt \tag{5-22}$$

From the function $v(t)$, we define its autocorrelation function as

$$C_v(\tau) = \langle v(t)v(t+\tau) \rangle = \frac{1}{T} \int_{-T/2}^{T/2} v(t)v(t+\tau)dt \tag{5-23}$$

Substituting $v(t)$ in Eq. (5-21) into Eq. (5-23), we obtain

$$C_v(\tau) = \frac{1}{T} \int_{-\infty}^{\infty} \int_{-\infty}^{\infty} \int_{-T/2}^{T/2} d\omega d\omega' dt V_T(\omega)V_T(\omega')e^{j(\omega+\omega')t}e^{j\omega\tau} \tag{5-24}$$

In the limit of $T \rightarrow \infty$,

$$\lim_{T \to \infty} \int_{-T/2}^{T/2} dt \ e^{j(\omega + \omega')t} = 2\pi\delta(\omega + \omega')$$

Therefore

$$C_v(\tau) = \lim_{T \to \infty} \frac{2\pi}{T} \int_{-\infty}^{\infty} \int_{-\infty}^{\infty} V_T(\omega')V_T(\omega)\delta(\omega + \omega') \ e^{j\omega\tau} d\omega d\omega'$$

$$= \lim_{T \to \infty} \frac{2\pi}{T} \int_{-\infty}^{\infty} V_T(-\omega)V_T(\omega) \ e^{j\omega\tau} d\omega \qquad (5\text{-}25)$$

If the function $v(t)$ is real, then according to Eq.(5-22),

$$V_T(-\omega) = V_T^*(\omega) \qquad (5\text{-}26)$$

Thus Eq. (5-25) becomes

$$C_v(\tau) = \lim_{T \to \infty} \frac{1}{2} \int_{-\infty}^{\infty} \frac{4\pi |V_T(\omega)|^2}{T} e^{j\omega\tau} d\omega \qquad (5\text{-}27)$$

where $\displaystyle \lim_{T \to \infty} \frac{4\pi |V_T(\omega)|^2}{T}$ is called spectral density function and is defined as $S_v(\omega)$.[6, 16]

$$S_v(\omega) = \lim_{T \to \infty} \frac{4\pi |V_T(\omega)|^2}{T} \qquad (5\text{-}28)$$

$S_v(\omega)$ represents the distribution of the power over the spectral domain. The total power is given by the integration of the spectral density function over the entire frequency domain, which is equal to the integration of the instantaneous power $v^2(t)$ over the time domain.

Eq. (5-27) states that the autocorrelation function is the Fourier transform of the spectral density function. In fact, the two functions form a Fourier-transformed pair.

$$C_v(\tau) = \frac{1}{2}\int_{-\infty}^{\infty} S_v(\omega)e^{j\omega\tau}d\omega \tag{5-29}$$

$$S_v(\omega) = \frac{1}{\pi}\int_{-\infty}^{\infty} C_v(\tau)e^{j\omega\tau}d\tau \tag{5-30}$$

The relation between the autocorrelation function and the spectral density function is called Wiener-Khintchine theorem.[6] Usually the power distribution over the spectrum is a very important piece of information for signals. This theory tells us that the spectral density function can be obtained either by the direct measurement of the power distribution or by the measurement of the autocorrelation function. In some cases, it is easier to measure $C_v(\tau)$ than $S_v(\omega)$. Eq.(5-30) is then used to obtain the power distribution.

3.2   Shot Noise

Shot noise [18] is the natural result of discrete signals. There are two points of view to look at the shot noise. First, because light has a particle nature, the light wave actually consists of discrete events of photons. Second, in the photodetector, the generated electronic signals are in the form of many electrons (or holes), which are discrete signals as well. In the detection of optical signals, the discrete photons are converted by a photodetector to the discrete electrons or holes, so the first viewpoint can also be equivalently treated as discrete electronic signals. In short, shot noise can be viewed as being due to the discrete nature of electronic signals that consist of many short pulses of current, each one being caused by an electron.

The total electronic signal expressed in the form of current is therefore the summation of many individual events of electron transit. Each one occurs at random time t_i.

$$i_T(t) = \sum_{i=1}^{N_T} i_e(t-t_i) \tag{5-31}$$

where N_T is the total number of events, here the number of transit electrons, occurring in the time duration T. $i_e(t-t_i)$ represents an individual signal expressed in the current resulting from the emission of an electron at t_i.

Taking the Fourier transform of the total electronic signal $i_T(t)$ expressed in Eq. (5-31), we obtain

$$I_T(\omega) = \sum_{i=1}^{N_T} I_i(\omega) \tag{5-32}$$

where $I_i(\omega)$ is the Fourier transform of the individual signal $i_e(t - t_i)$.

$$I_i(\omega) = \frac{1}{2\pi} \int_{-\infty}^{\infty} i_e(t - t_i)\, e^{-j\omega t}\, dt$$

$$= e^{-j\omega t_i} I(\omega) \tag{5-33}$$

where $I(\omega)$ is the Fourier transform of signal $i_e(t)$. $i_e(t)$ is the current resulting from the emission of an electron at time $t = 0$. This current flows only when the carrier is in transit. The transit time $\tau = d/v$, where d is the distance for the carrier to transit and v is the velocity of carrier. The change of velocity during the carrier transit is neglected.

$$I(\omega) = \frac{1}{2\pi} \int_0^\tau i(t) e^{-j\omega t}\, dt = \frac{ev}{2\pi d} \int_0^\tau e^{-j\omega t}\, dt$$

$$= \frac{e}{2\pi} \frac{\sin(\omega\tau/2)}{(\omega\tau/2)} e^{-j(\omega\tau/2)} \tag{5-34}$$

The detection of a photon results in a current pulse with its frequency components given by Eq. (5-34). The transit time τ is usually very short. For example, in the PIN-photodiode with the intrinsic layer of 1 μm, the transit time is only around 10 ps, which is usually much shorter than the interested duration of signals. Thus we take the limit of $\tau \to 0$. Then

$$I(\omega) = \frac{e}{2\pi} \tag{5-35}$$

From Eq. (5-32), the Fourier transform of the total electronic signal $i_T(t)$ then becomes

$$I_T(\omega) = \sum_{i=1}^{N_T} I_i(\omega) = \frac{e}{2\pi} \sum_{i=1}^{N_T} e^{-j\omega t_i} \tag{5-36}$$

According to Eq. (5-28), the spectral density function of the total electronic signal is given by

$$S(\omega) = \lim_{T \to \infty} \frac{4\pi |I_T(\omega)|^2}{T} \tag{5-37}$$

where

$$|I_T(\omega)|^2 = (\frac{e}{2\pi})^2 (\sum_{i=1}^{N_T} e^{-j\omega t_i})(\sum_{k=1}^{N_T} e^{j\omega t_k})$$

$$= (\frac{e}{2\pi})^2 [N_T + \sum_{i \neq k}^{N_T} \sum_{k=1}^{N_T} e^{j\omega(t_k - t_i)}] \tag{5-38}$$

As we consider the randomness of the arrival time t_i and the large number of events, the average of the second term in the right hand side of Eq. (5-38) will be near to zero. Therefore,

$$\overline{|I_T(\omega)|^2} = (\frac{e}{2\pi})^2 \overline{N_T} = (\frac{e}{2\pi})^2 \overline{N} \, T \tag{5-39}$$

where $\overline{N}$ is the average number of events that occur per unit time. Thus the spectral density function is given by

$$S(\omega) = \frac{e^2}{\pi} \overline{N} \tag{5-40}$$

It is more common to use the spectral density function in terms of ν, where $\nu = \omega/2\pi$. Taking the fact that $S_\nu(\omega)d\omega = S_\nu(\nu)d\nu$, we have $S_\nu(\nu) = 2\pi S_\nu(\omega)$. On the other hand, the many current pulses give rise to an average DC current

$$\overline{I} = e\overline{N}$$

Then the spectral density function becomes

$$S(\nu) = 2e\overline{I} \tag{5-41}$$

Therefore, even when we measure the unmodulated lightwave using a photodetector, we obtain two parts at the output. The first one is the generation of a DC current. The second one is the undesired signal, called shot noise. It is due to the discrete nature of signals. The shot noise has equal power at all frequency components. If the electronic circuits after the photodetector only handle the frequency band from ν to $\nu + \Delta\nu$, the mean square current amplitude of the shot noise that accompanies with the signal is given by

$$\overline{i_N^2}(\nu) = S(\nu)\Delta\nu = 2e\overline{I}\Delta\nu \qquad (5\text{-}42)$$

For APDs, the shot noise also goes through the multiplication process. In addition to the amplified shot noise from the primary photocurrent generated before avalanche process, excess noise is generated during the multiplication process. An excess noise factor $F(M)$ is used to describe the additional noise. Therefore the shot noise for APDs is given by

$$\overline{i_N^2}(\nu) = 2e\overline{I_p}M^2F(M)\Delta\nu = 2e\overline{I_M}MF(M)\Delta\nu \qquad (5\text{-}43)$$

where I_P is the unmultiplied photocurrent, I_M is the multiplied output current, and M is the multiplication factor.

3.3 Thermal Noise

Thermal noise is also called Johnson noise[19] or Nyquist noise.[20] Thermal noise is caused by the thermal agitation of charge carriers passing through a resistor. For simplicity of derivation, we assume that the resistor has a volume $V = Ad$, where A is the cross section of the resistor and d is the length along the direction that current flows. In this resistor, there are N free electrons and N ions of positive charge per unit volume.

Electrons in the resistor move randomly. The average kinetic energy for every electron is

$$\overline{E} = \frac{3}{2}kT = \frac{1}{2}m(\overline{v_x^2} + \overline{v_y^2} + \overline{v_z^2}) \qquad (5\text{-}44)$$

where $\overline{v_x^2} = \overline{v_y^2} = \overline{v_z^2} = kT/m$. The resistance is caused by the electron-electron, electron-ion, and electron-phonon scattering. The scattering process is usually characterized by the mean scattering time τ_0, which means the average time between two successive collisions. The scattering mechanisms then give rise to the resistance R. [21, 22]

$$R = \frac{md(1 + \omega^2 \tau_0^2)}{Ne^2 \tau_0 A} \tag{5-45}$$

The motion of one electron between two successive collisions leads to the current pulse $i_e(t)$.

$$i_e(t) = \begin{cases} ev_x / d & 0 \le t \le \tau \\ 0 & otherwise \end{cases} \tag{5-46}$$

Here we assume that the electron moves along the x-direction in that period. We assume that the velocity is constant. τ is the scattering time of that event. Taking the Fourier transform of the current pulse $i_e(t)$, we obtain

$$I_e(\omega, \tau, v_x) = \frac{1}{2\pi} \int_0^\tau i_e(t) e^{-j\omega t} dt = \frac{(1/2\pi)ev_x}{-j\omega d}(e^{-j\omega \tau} - 1) \tag{5-47}$$

Because the spectral density function is proportional to the absolute square of the above quantity, we do the following evaluation

$$\left| I_e(\omega, \tau, v_x) \right|^2 = \frac{e^2 v_x^2}{4\pi^2 \omega^2 d^2}(2 - e^{-j\omega \tau} - e^{j\omega \tau}) \tag{5-48}$$

The above calculation is for one single event of scattering. The scattering time τ as well as the velocity v_x is different for every scattering event. The scattering time τ has the following probability function

$$g(\tau) = \frac{1}{\tau_0} e^{-\tau/\tau_0} \tag{5-49}$$

The velocity follows the Maxwell distribution $f(v_x)$. Taking the average over many events using those probability functions, we obtain the quantity independent of the individual scattering time and the individual velocity

$$\overline{|I_e(\omega)|^2} = \frac{2e^2\tau_0^2\overline{v_x^2}}{4\pi^2d^2(1+\omega^2\tau_0^2)} \qquad (5\text{-}50)$$

Now we have the quantity dependent on the mean scattering time and the mean-square velocity. Using $\overline{v_x^2} = \overline{v_y^2} = \overline{v_z^2} = kT/m$, we then obtain

$$\overline{|I_e(\omega)|^2} = \frac{2e^2\tau_0^2kT}{4\pi^2md^2(1+\omega^2\tau_0^2)} \qquad (5\text{-}51)$$

In the resistor, there are NV electrons. On average, each electron goes through a scattering event for every mean scattering time τ_0, so the number of scattering events per second is given by

$$N = \frac{NV}{\tau_0} \qquad (5\text{-}52)$$

From Eq.(5-28), we have

$$S(v) = 8\pi^2\overline{N}\overline{|I_e(\omega)|^2} = \frac{4NVe^2\tau_0kT}{md^2(1+\omega^2\tau_0^2)}$$

$$= \frac{4Ne^2\tau_0AkT}{md(1+\omega^2\tau_0^2)} \qquad (5\text{-}53)$$

Using the resistance formula given by Eq.(5-45), we obtain

$$S(v) = \frac{4kT}{R(v)} \qquad (5\text{-}54)$$

Again, if the circuits handle only a bandwidth of Δv and the resistance is constant within this bandwidth, then the mean-square current amplitude of the thermal noise that accompanies with the signal is given by

$$\overline{i_N^2}(v) = S(v)\Delta v = \frac{4kT}{R}\Delta v \qquad (5\text{-}55)$$

The above derivation is based on the current point of view. It can also be derived using the voltage point of view. The mean-square voltage amplitude of the thermal noise that accompanies with the signal is given then by

$$\overline{v_N^2}(v) = 4kTR\Delta v \tag{5-56}$$

3.3   Dark Current Noise and Background-Illumination Noise

Dark current is the current generated in the photodiode even without the illumination of light. It is equivalent to the reverse-bias current of a diode. The dark current strongly depends on the type of semiconductors, operating temperature, and bias voltage. It is typically proportional to the term exp (-Eg/kT). Dark current is in the order of 100 pA for Si PIN-photodiodes and 10 pA for Si APD photodiode. For photodiodes made of low bandgap semiconductors, dark current increases. For InGaAs-InP PIN-photodiodes and APDs, the dark current is around 2 nA- 5 nA. In Ge APDs, the dark current is around 100 nA.

The dark current can be separated to two parts. One is due to the thermally generated electron-hole pairs in the photodiode. For APDs, this current also goes through multiplication process, so this part of dark current is amplified. The other part of dark current is caused by the surface recombination due to defects or surface states, etc. It is already at the surface of the device, so no multiplication occurs.

Background-illumination noise is caused by the light that is not a part of the transmitted optical signals. The measurement environment is usually not in a dark room. The ambient light contributes a lot to the background illumination. If the photodiode is not isolated from the background radiation, the appearance of background-illumination noise is inevitable. It is given by

$$I_{BK} = \eta_{qe} \bullet e \bullet \frac{P_{BK}}{hv} \tag{5-57}$$

Because the dark current noise and the background-illumination noise are both discrete in nature, they will also give rise to shot noise. As a result, shot noise is produced by three sources: actual optical signals, dark current, and background illumination.

3.4 Signal-to-Noise Ratio

After we realize the sources of noise, we are able to evaluate the signal-to-noise ratio, which is simply expressed as S/N ratio. In optical detection, the signal level must not be less than the noise level. That is the S/N ratio ≥ 1. The S/N ratio for different photodiodes will be discussed in the following.

3.4.1 S/N Ratio for PIN-photodiode

The photocurrent corresponding to the optical signals is given by

$$I_{pc} = \eta_{qe} \bullet e \bullet \frac{P_{abs}}{h\nu} = R_0 P_{abs} \tag{5-58}$$

where R_0 is the responsivity of the PIN-photodiode. The mean square current amplitude of signals is thus

$$S = R_0^2 P_{abs}^2 \tag{5-59}$$

The noise of the PIN-photodiode includes
- shot noise;
- thermal noise:
- dark current noise;
- background-illumination noise.

Shot noise is produced by actual optical signals, dark current, and background illumination. Thus it is given by

$$\overline{i_N^2}(\nu) = 2e(I_{pc} + I_{dark} + I_{BK})\Delta\nu$$

$$= 2e(R_0 P_{abs} + I_{dark} + R_0 P_{BK})\Delta\nu \tag{5-60}$$

Thermal noise is given by Eq. (5-55). Therefore the total noise is as follows

$$Noise = \overline{i_N^2}(\nu) \; (shot) + \overline{i_N^2}(\nu) \; (thermal) + R_0^2 P_{BK}^2 B_{r1} + I_{dark}^2 B_{r2}$$

$$= 2e(R_0 P_{abs} + I_{dark} + R_0 P_{BK} + \frac{4kT}{R})\Delta\nu + R_0^2 P_{BK}^2 B_{r1} + I_{dark}^2 B_{r2} \tag{5-61}$$

where B_{r1} and B_{r2} are the bandwidth ratios, respectively, for the background-illumination noise and dark current noise that enter the bandwidth of interest. Because the background-illumination and dark current may spread over a spectral domain beyond the bandwidth that the electronic circuits handle, not all of them accompany the signals. Therefore, only that portion of noise within the bandwidth is taken into account.

The S/N ratio is thus given in the following

$$\frac{S}{N} = \frac{R_0^2 P_{abs}^2}{2e(R_0 P_{abs} + I_{dark} + R_0 P_{BK} + \frac{4kT}{R})\Delta v + R_0^2 P_{BK}^2 B_{r1} + I_{dark}^2 B_{r2}}$$

$$(5\text{-}62)$$

Not all of the noise sources have the same importance. Some may dominate over others, depending on the detection condition. When the power of the optical signals is high, the shot noise produced by the optical signals is much larger than other noises. The S/N ratio becomes

$$\frac{S}{N} = \frac{R_0^2 P_{abs}^2}{2e R_0 P_{abs} \Delta v} = \frac{R_0 P_{abs}}{2e\Delta v} \qquad (5\text{-}63)$$

When the power of the optical signals and the background illumination is low, thermal noise and dark current dominate. Then the S/N ratio becomes

$$\frac{S}{N} = \frac{R_0^2 P_{abs}^2}{2e(I_{dark} + \frac{4kT}{R})\Delta v + I_{dark}^2 B_{r2}} \qquad (5\text{-}64)$$

The dark current of Si PIN-photodiodes and InGaAs-InP PIN-photodiodes can often be neglected, so only thermal noise is taken into account and the S/N ratio becomes

$$\frac{S}{N} = \frac{R_0^2 P_{abs}^2}{4kT\Delta v / R} \qquad (5\text{-}65)$$

3.4.2   S/N Ratio for APD Photodiode

For APD photodiodes, the photocurrent corresponding to the optical signals is given by

$$I_M = I_p M = R_0 P_{abs} M \tag{5-66}$$

where I_P is the unmultiplied photocurrent, I_M is the multiplied output current, and M is the multiplication factor of the APD.

$$S = I_p^2 M^2 = R_0^2 P_{abs}^2 M^2 \tag{5-67}$$

The noise of the APD also includes
– shot noise;
– thermal noise;
– dark current noise;
– background-illumination noise.

Shot noise is produced by actual optical signals, dark current, and background illumination. In addition, the dark current has two parts that contribute to the shot noise in different ways. As described before, one part is amplified in the avalanche process and the other is not. The shot noise produced by the optical signal is given by Eq. (5-43), while the shot noise produced by the background illumination can be obtained from the same formula with the photocurrent of optical signals replaced with the photocurrent of background illumination. Thus the shot noise is given by

$$\overline{i_N^2}(v) = 2e(\overline{I_p}M^2 F(M) + I_{dk1}M^2 F(M) + I_{dk2} + I_{BK}M^2 F(M))\Delta v \tag{5-68}$$

where I_{dk1} and I_{dk2} are the two parts of dark current.

Thermal noise is again given by Eq. (5-55). Therefore the total noise is as follows

$$Noise = \overline{i_N^2}(v) \ (shot) + \overline{i_N^2}(v) \ (thermal) + I_{BK}^2 M^2 B_{r1}$$
$$+ (I_{dk1}^2 M^2 + I_{dk2}^2) B_{r2} \tag{5-69}$$

The shot noise is given by Eq. (5-68) and the thermal noise is given by Eq. (5-55). Also, we only take into account the background-illumination noise and dark current noise that enter the bandwidth of interest.

The S/N ratio is thus given in the following

$$\frac{S}{N} = \frac{R_0^2 P_{abs}^2 M^2}{i_N^2(v) \ (shot) + i_N^2(v) \ (thermal) + I_{BK}^2 M^2 B_{r1} + (I_{dk1}^2 M^2 + I_{dk2}^2) B_{r2}}$$

(5-70)

Again, not all of the noise components have the same importance. For high-power optical signals, the shot noise due to the optical signals dominates. The S/N ratio then becomes

$$\frac{S}{N} = \frac{R_0^2 P_{abs}^2 M^2}{i_N^2(v) \ (shot)} = \frac{R_0^2 P_{abs}^2 M^2}{2eR_0 P_{abs} M^2 F(M) \Delta v} = \frac{R_0 P_{abs}}{2eF(M) \Delta v}$$

(5-71)

When the power of the optical signals and the back ground illumination is low, thermal noise and dark current will be dominant. Then the S/N ratio becomes

$$\frac{S}{N} = \frac{R_0^2 P_{abs}^2 M^2}{4kT\Delta v / R + 2e(I_{dk1} M^2 F(M) + I_{dk2})\Delta v + (I_{dk1}^2 M^2 + I_{dk2}^2) B_{r2}}$$

(5-72)

As the dark current is further neglected, then only thermal noise is taken into account. The S/N ratio becomes

$$\frac{S}{N} = \frac{R_0^2 P_{abs}^2 M^2}{4kT\Delta v / R}$$

(5-73)

Comparing Eq. (5-65) for PIN-photodiode and Eq. (5-73) for APD, we can see that the S/N ratio for APD is improved for M^2. This is the case when the power of the optical signals is low and the thermal noise dominates over other types of noises. On the other hand, if the power of the optical signals is high, then the comparison between Eq. (5-63) and Eq. (5-71) shows that the PIN-photodiode has better S/N ratio for a factor of $F(M)$ because the factor $F(M)$ is approximately equal to M^x. x is typically 0.4 for Si, 0.7 for InGaAs and 1.0 for Ge APD. This means that APDs are not necessarily better than PIN-photodiodes for the detection of optical signals.

NOTES

1. For general discussion on photodetectors used in optical communication, please refer to Refs. 1-3.
2. Readers that want to know detail of the p-n junction are referred to Ref. 5.
3. The classic reference to the discussion on shot noise is in Ref. 18.
4. Early discussions on the thermal noise are given in Ref. 19-20.

REFERENCES

1. Pearsall T., "Photodetectors for Communication by Optical Fibers." In *Optical Fiber Communications*. M. J. Howes and D. V. Morgan, eds., Chapter 6, New York: John Wiley & Sons, Inc., 1980.
2. Powers, J. P., *An Introduction to Fiber Optic System*. Aksen Associates Inc., 1988.
3. Hunsperger, Robert G., *Photonic Devices and System*. Marcel Dekker Inc., 1994.
4. Wilson, John and Hawkes, John, *Optoelectronics: an Introduction*. 3/ed, Prentice Hall, 1998.
5. Sze, S. M., *Semiconductor Devices: Physics and Technology*. John Wiley & Sons, Inc., 1985.
6. Yariv, Amnon, *Optical Electronics in Modern Communications*. 5/ed, Oxford University Press, 1997.
7. Franz, J. H. and Jain, V. K., *Optical communications: components and system*. CRC Press, 2000.
8. Bowers, J. E. and Burrus, C. A., Ultrawide-band long-wavelength p-i-n photodetectors. Journal of Lightwave Technology 1987; 14: 1339-1350.
9. Kato, K., Hata, S., Kawano, K., Yoshida, J., and Kozen, A., A high-efficiency 50 GHz InGaAs multimode waveguide photodetector. IEEE Journal of Quantum Electronics 1992; 28: 2728-2735.
10. Hietala, V. M., Vawter, G. A., Brennan, T. M., and Hammons, B. E., Travelling-wave photodetectors for high-power, large-bandwidth applications. IEEE Transaction MTT 1995; 43: 2291-2298.
11. Streetman, B. G., *Solid State Electronic Devices*. 2/ed, Prentice Hall, 1980.
12. Stillman G. E. and Wolfe, C. M., Avalanche photodiodes. In *Semiconductors and semimetals*. Willardson R.K. and Beer A.C., eds., vol. 12, Chapter 5, New York: Academic Press, 1977.
13. Wilson, John and Hawkes, John, *Optoelectronics*. Prentice Hall, 1989.
14. Smith, R. G., Photodetectors for fiber transmission system. Proceedings of IEEE 1980; 68: 1247-1253.
15. Keiser, Gerd, *Optical Fiber Communication*. 2/ed, McGraw-Hill, 1992.
16. Verdeyen, J. T., Laser Electronics. 3/ed, Prentice Hall, 1995.
17. Davenport, W.B. and Root, W. L., *An Introduction to the Theory of Random Signals and Noise*. McGraw-Hill, 1958.
18. Rice, S. O., Mathematical analysis of random noise. Bell Syst. Tech. J. 1944; 23: 282-332. 1945; 24: 46- 156.
19. Johnson, J. B., Thermal agitation of electricity in conductors. Phys. Rev. 1928; 32: 97-109.
20. Nyquist, H., Thermal agitation of electric charge in conductors, Phys. Rev. 1928; 32: 110-113.

21. Van der Ziel, A., Solid State Physical Electronics, Holonyak, Nick Jr. ed., Prentice Hall, 1968.
22. Reif, F., *Fundamentals of Statistical and Thermal Physics*. McGraw-Hill, 1965.

Chapter 6

PRINCIPLES OF OPTICS FOR PASSIVE COMPONENTS

1. INTRODUCTION

In a fiber optic communication system, many different optical components are used. In previous chapters, we discussed active components, which are used for three purposes:
1. converting electrical signals to optical signals;
2. amplifying optical signals transmitted in the fiber;
3. converting optical signals back to electrical signals.

Passive components have various functions much more diverse than active components although most require no external power to perform those functions. Some examples of those components are shown in the following. Block diagrams are used to denote their functions.

1.1 Combiner

The function of a combiner is shown in Fig.6-1. Its purpose is to combine two or more inputs of optical signals from different paths into one output so that they can be delivered to the same destination.

Figure 6-1. Function of a combiner denoted by the block diagram.

1.2   Splitter

The function of a splitter is shown in Fig.6-2. Its function is just the opposite to that of a combiner. Its purpose is to split optical signals from one path to two or more outputs so that the same signals can be delivered to the different destinations at the same time.

Figure 6-2. Function of a splitter denoted by the block diagram.

1.3   N x N Coupler

The functions of combining and splitting can be put together in a single device. The number of inputs and outputs is also not limited to two. Fig. 6-3 shows the block diagram of an N x N coupler. In this device, input 1 can be split into N outputs. N inputs can also be combined and delivered to any output. The ratio of power combining or splitting can be mathematically expressed in the following matrix formula.

$$\begin{pmatrix} P_{o1} \\ P_{o2} \\ P_{o3} \\ \vdots \\ P_{oN} \end{pmatrix} = \begin{pmatrix} S_{11} & S_{12} & S_{13} & \cdots & S_{1N} \\ S_{21} & S_{22} & S_{23} & \cdots & S_{2N} \\ S_{31} & S_{32} & S_{33} & \cdots & S_{3N} \\ \vdots & \vdots & \vdots & \ddots & \vdots \\ S_{N1} & S_{N2} & S_{n3} & \cdots & S_{NN} \end{pmatrix} \begin{pmatrix} P_{i1} \\ P_{i2} \\ P_{i3} \\ \vdots \\ P_{iN} \end{pmatrix} \qquad (6\text{-}1)$$

where P_{ik} ($k = 1, 2, \ldots, N$) is the power of input k and P_{om} ($m = 1, 2, \ldots, N$) is the power of output m.

Figure 6-3. Function of an N x N coupler denoted by the block diagram.

1.4 Wavelength Division Multiplexer and Demultiplexer (WDM)

In WDM communication systems, different channels of optical signals are assigned to different wavelengths. Those many channels are combined and delivered in a single optical fiber. A wavelength division multiplexer (WDM) is used to combine two or more inputs of optical signals at different wavelengths and from different paths to one output in order to deliver them simultaneously to one single optical fiber. In contrast, a wavelength division demultiplexer is used to separate optical signals at different wavelengths from one single optical fiber to many outputs according to their wavelengths. Different outputs are then delivered to different destinations. Their block diagrams are shown in Fig. 6-4 (a) and (b), respectively.

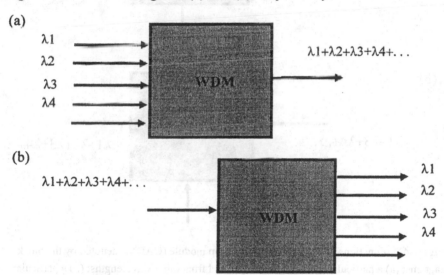

Figure 6-4. Functions of (a) wavelength division multiplexer and (b) wavelength division demultiplexer denoted by the block diagram.

1.5 Optical Add and Drop Module

In WDM communication systems, it is possible that a new channel of optical signals, assigned to a wavelength different from those existing in current system, is added in a later time. An optical add and drop module (OADM) is used to add this new channel to the current system. In this way, the WDM system has the flexibility to expand its communications capacity or reduce its coverage range to reduce cost if necessary. Thus an OADM is used to separate optical signals at one wavelength from those at other wavelengths so that a set of optical signals at a particular wavelength can be delivered to a destination different from others. In addition, it can serve to add optical signals from a particular source (assigned to a particular wavelength) to the system, so information from this special source can be delivered to other places using the same system. Its function is shown in Fig. 6-5. Fig. 6-5(a) shows that signals at wavelength $\lambda 2$ is dropped from the system, while Fig. 6-5(b) shows that signals at wavelength $\lambda 2$ is added to the system.

(a)

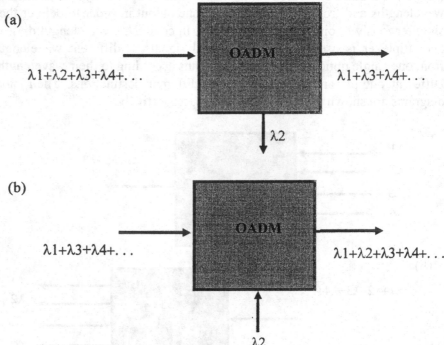

(b)

Figure 6-5. Functions of an optical add and drop module (OADM) denoted by the block diagrams: (a) a particular wavelength is separated from other wavelengths; (b) a particular wavelength is added to other wavelengths transmitting in the same communication system.

1.6   Discussion of Passive Components

The components introduced above are only a few examples. There are more passive components than those mentioned. Although there are many types of passive components providing many different functions, their principles of operations are not completely dissimilar. In addition, several principles of optics may be combined together in a single type of component. Furthermore, one particular principle of optics may be applied to several types of passive components. Therefore, it is better to discuss those individual optical principles before we study specific components.

Those principles or phenomena that are most often encountered in passive components will be discussed in this chapter. They include
- Fabry-Perot resonance due to parallel interfaces;
- transmission and reflection from multi-layer coatings;
- principle of diffraction grating;
- waveguide coupling;
- interference of Mach-Zehnder interferometer;
- Birefringence;
- Optical activity.

2. FABRY-PEROT RESONANCE DUE TO PARALLEL INTERFACES

As light passes through an optical component, it will see at least two interfaces lying between the material it is made of and the outside environment. For example, when light passes through a piece of glass placed in the air, the light will first enter the glass by penetrating the first interface, then leave the glass and return to the air by penetrating the second interface. As we discussed in Chapter 1, at the interface, light has transmission and reflection. The first interface will cause part of light to reflect and other part of light to transmit. The transmitted light meets the second interface again has transmission and reflection. Thus, the second interface also causes this light to partially reflect and transmit. The situation is illustrated in Fig. 6-6.

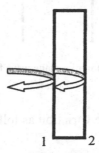

Figure 6-6. Light passing an optical component meets two interfaces.

The situation is actually quite complicated. For the light transmitted after the first interface, as it propagates to the second interface, it will be partially reflected. This reflected light will propagate to the first interface, which again causes the light to be partially reflected and transmitted. The reflected light will then meet the second interface. The situation goes on, so some portion of light bounces between the two interfaces. It is thus necessary to analyze the overall effect due to many instances of reflection and transmission at the two interfaces.

2.1   Analysis of Fabry-Perot Resonance

The reflection and transmission due to the two parallel interfaces can be analyzed using the following approach. Similar to the analysis of the reflection and transmission from a single interface in Chapter 1, light is treated as an electromagnetic wave. Fig. 6-7 shows the reflection and transmission due to the two interfaces. The interfaces are perpendicular to the x-axis. One is at x = 0, called the first interface, and the other is at x = d, called the second interface. A plane wave is propagating from the left (x < 0) and incident on the first interface. Its direction of propagation is on the x-z plane. This direction is the same as the wave vector $\bar{k}_1$ in the region x < 0. The incident angle, defined as the angle between the $\bar{k}_1$ vector and the normal of the interface (x-axis), is θ_1.

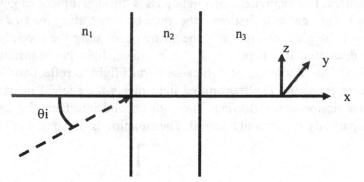

Figure 6-7. Reflection and transmission of a plane wave from two interfaces parallel to the y-z plane.

The refractive index has the variation as follows.

$$n(x) = \begin{cases} n_1 & x < 0 \\ n_2 & 0 < x < d \\ n_3 & x > d \end{cases} \qquad (6\text{-}2)$$

Therefore, the first interface is between the two materials with the refractive indices n_1 and n_2, respectively, while the second interface is between the two materials with the refractive indices n_2 and n_3, respectively.

Because the variation of refractive index is only along the x-axis, the electric field also has the variation of amplitude along the x-direction only. In the z-direction, its variation is only at the phase term.

$$\bar{E} = \bar{E}(x)e^{j(\omega t - \beta z)} \tag{6-3}$$

where β is the component of the wave vector $\bar{k}_i$ along the z–direction.

$$\bar{k}_i = k_{ix}\hat{x} + \beta\hat{z} \tag{6-4}$$

β is the same for all of the three regions.

The wave vector $\bar{k}_i$ has the magnitude equal to $\dfrac{n_i\omega}{c}$. For example, in medium 1 (x<0), the refractive index is n_1, so $\left|\bar{k}_1\right| = \dfrac{n_1\omega}{c}$. Similarly, the magnitude of wave vector in medium 2 and 3 are $\dfrac{n_2\omega}{c}$ and $\dfrac{n_3\omega}{c}$, respectively. The z-component of the wave vector $\bar{k}_i$ is the same for all three regions because the index has no variation along the z direction. $\left|\bar{k}_1\right|\sin\theta_1 = \left|\bar{k}_2\right|\sin\theta_2 = \left|\bar{k}_3\right|\sin\theta_3$, so we have the Snell's law for the two interfaces.

$$n_1 \sin\theta_1 = n_2 \sin\theta_2 = n_3 \sin\theta_3 \tag{6-5}$$

where θ_i is the angle between the wave vector $\bar{k}_i$ in medium i with the refractive index n_i and the normal direction (x-axis). In addition, we obtain the x-component of the wave vector $\bar{k}_i$ in medium i given by

$$k_{ix} = \left[(\frac{n_i\omega}{c})^2 - \beta^2\right]^{1/2} = \frac{\omega}{c}n_i\cos\theta_i, \, i = 1, 2, 3 \tag{6-6}$$

In Chapter 1, we derived the reflection and transmission of a single interface. In the derivation, we discovered that the incident wave can be decomposed into a TE wave and a TM wave, which each have different reflection and transmission coefficients. That is, regardless of reflection or transmission, a TE wave does not change its polarization by the interface and neither does the TM mode. As a result, the TE wave and TM wave can be manipulated separately.

2.1.1   Fabry-Perot Resonance for TE Wave

Let us look at the TE wave first. For the TE wave, the electric field has only the y-component. According to Eq.(6-3), the electric field can be written as

$$\bar{E} = E_y(x)e^{j(\omega t - \beta z)}\,\hat{y} \qquad\qquad (6\text{-}7)$$

Consider the wave incident from the region $x < 0$. Some portion of this wave is reflected back to the medium one due to the two interfaces, so the wave has two parts. One propagates toward the +x direction and the other propagates toward −x direction. In medium two, some portion of wave bounces between the two interfaces, so the wave also has two parts. One is toward the +x direction and the other is toward −x direction. In medium three, there is no wave reflected by other interface, so only the wave propagating toward +x direction exists. Therefore, we have

$$E_y(x) = \begin{cases} Ae^{-jk_{1x}x} + Be^{jk_{1x}x} & x < 0 \\ Ce^{-jk_{2x}x} + Be^{jk_{2x}x} & 0 < x < d \\ Fe^{-jk_{3x}x} & d < x \end{cases} \qquad (6\text{-}8)$$

The magnetic field can be obtained from Maxwell's equations, $\bar{H} = \dfrac{\bar{k} \times \bar{E}}{\omega\mu}$. Then we obtain

$$H_z = \begin{cases} \dfrac{k_1 x}{\omega\mu}\left(Ae^{-jk_{1x}x} - Be^{jk_{1x}x}\right) & x < 0 \\[2mm] \dfrac{k_{2x}}{\omega\mu}\left(Ce^{-jk_{2x}x} - De^{jk_{2x}x}\right) & 0 < x < d \\[2mm] \dfrac{k_{3x}}{\omega\mu}Fe^{-jk_{3x}x(x-d)} & d < x \end{cases} \qquad (6\text{-}9)$$

According to Eq.(1-24a) and Eq. (1-24b), the tangential components of E field and H field have to be continuous at the interface. Therefore, E_y and H_z should be continuous at $x = 0$ and $x = d$. Then we will have four equations to solve for the coefficients A, B, C, D, and F.

$$A + B = C + D \qquad\qquad (6\text{-}10a)$$

$$k_{1x}(A - B) = k_{2x}(C - D) \tag{6-10b}$$

$$Ce^{-jk_{2x}d} + De^{jk_{2x}d} = F \tag{6-10c}$$

$$k_{2x}\left(Ce^{-jk_{2x}d} - De^{jk_{2x}d}\right) = k_{3x}F \tag{6-10d}$$

With some mathematical manipulation, we obtain the reflection coefficient (r) and the transmission coefficient (t) as follows.

$$r \equiv \frac{B}{A} = \frac{r_{12} + r_{23}e^{-2j\phi}}{1 + r_{12}r_{23}e^{-2j\phi}} \tag{6-11}$$

$$t = \frac{F}{A} = \frac{t_{12}t_{23}e^{-j\phi}}{1 + r_{12}r_{23}e^{-2j\phi}} \tag{6-12}$$

where ϕ is the phase that the wave experiences in medium two.

$$\phi = k_{2x}d = \frac{2\pi d}{\lambda}n_2\cos\theta_2 \tag{6-13}$$

r_{12} and t_{12} are the reflection coefficient and the transmission coefficient respectively, for the instance when the TE wave propagates from medium one to medium two and meets the first interface. The influence of the second interface is not included in r_{12} and t_{12}. Similarly, r_{23} and t_{23} are the reflection coefficient and the transmission coefficient of TE wave when it meets the second interface. As before, the influence of the first interface is not included in r_{23} and t_{23}. Those coefficients have already given in Chapter 1. The reflection and transmission due the individual interface is shown in Fig.6-8(a) for the first interface and Fig.6-8(b) for the second interface. Modifying Eqs. (1-39a) and (1-39b) to the case shown in Fig. 6-8, we obtain

$$r_{12} = \frac{n_1\cos\theta_1 - n_2\cos\theta_2}{n_1\cos\theta_1 + n_2\cos\theta_2} \quad \text{(TE)} \tag{6-14a}$$

$$t_{12} = \frac{2n_1\cos\theta_1}{n_1\cos\theta_1 + n_2\cos\theta_2} \quad \text{(TE)} \tag{6-14b}$$

$$r_{23} = \frac{n_2 \cos\theta_2 - n_3 \cos\theta_3}{n_2 \cos\theta_2 + n_3 \cos\theta_3} \quad \text{(TE)} \tag{6-15a}$$

$$t_{23} = \frac{2n_2 \cos\theta_2}{n_2 \cos\theta_2 + n_3 \cos\theta_3} \quad \text{(TE)} \tag{6-15b}$$

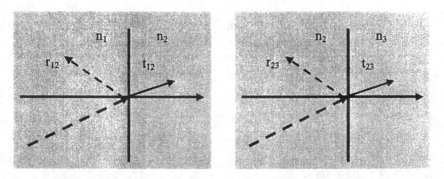

Figure 6-8. Reflection and transmission from an individual interface: (a) between n_1 and n_2; (b) between n_2 and n_3.

2.1.2   Fabry-Perot Resonance for TM Wave

For the TM wave, we can derive the reflection and transmission coefficients with similar mathematical manipulation. The reflection coefficient (r) and the transmission coefficient (t) have the same mathematical formula as Eq. (6-11) and Eq. (6-12), respectively. However, the coefficients r_{12}, t_{12}, r_{23} and t_{23} used in the formula are different from those given by Eqs.(6-14a), (6-14b), (6-15a), and (6-15b). They have to be modified for the TM wave. From Eqs. (1-38a) and (1-38b) for TM wave in Chapter 1, we obtain

$$r_{12} = -\frac{n_2 \cos\theta_1 - n_1 \cos\theta_2}{n_2 \cos\theta_1 + n_1 \cos\theta_2} \quad \text{(TM)} \tag{6-16a}$$

$$t_{12} = \frac{2n_1 \cos\theta_1}{n_2 \cos\theta_1 + n_1 \cos\theta_2} \quad \text{(TM)} \tag{6-16b}$$

$$r_{23} = -\frac{n_3 \cos\theta_2 - n_2 \cos\theta_3}{n_3 \cos\theta_2 + n_2 \cos\theta_3} \quad \text{(TM)} \tag{6-17a}$$

$$t_{23} = \frac{2n_2 \cos\theta_2}{n_3 \cos\theta_2 + n_2 \cos\theta_3} \quad \text{(TM)} \tag{6-17b}$$

2.2   An Example of Fabry-Perot Resonance

The effect of two parallel interfaces can be understood from a piece of glass placed in the air. In this case, n_1 and n_3 are the refractive index of air, $n_1 = n_3 = 1$. n_2 is the refractive index of glass, $n_2 = 1.5$. Assuming normal incidence of the wave on the glass. Substituting the values of refractive indices into Eqs. (6-11), (6-12), (6-14a), (6-14b), (6-15a), and (6-15b), we obtain

$$r = \frac{\rho\left(e^{-2j\phi} - 1\right)}{1 - \rho^2 e^{-2j\phi}} \tag{6-18}$$

$$t = \frac{\left(1 - \rho^2\right) e^{-j\phi}}{1 - \rho^2 e^{-2j\phi}} \tag{6-19}$$

where $\rho = (n_2-1)/(n_2+1) = 0.2$. $\phi = \dfrac{2\pi d}{\lambda} n_2 \cos\theta_2$. r and t are the ratios of the amplitude of the field. The reflectivity and transmission of the intensity or power is given by $R \equiv |r|^2$ and $T \equiv |t|^2$. We should have $R + T = 1$ when $n_1 = n_3$.

From Eqs.(6-18) and (6-19), we know that R and T are periodic functions of ϕ. Furthermore, ϕ is a function of wavelength, thickness of glass, and the angle θ_2. If the wavelength and the angle are fixed, then R and T are periodic functions of the thickness of glass. If the thickness of the glass and the angle are fixed, then R and T are periodic functions of the wavelength. The transmission of power (T) is plotted in Fig. 6-9 as a function of wavelength, showing the periodic variation of the transmission with the wavelength. For the glass, the peak transmission (T_{max}) is 1.0, while the minimum transmission (T_{min}) is 0.85. The neighboring peaks of transmission are separated with the spectral spacing $\Delta\lambda$.

$$\Delta\lambda = \frac{\lambda^2}{2n_2 d} \tag{6-20}$$

In comparison with Eq. (2-35), we find that this spectral spacing is the same as the mode spacing of a laser with the cavity formed by two mirrors. The variation of reflection with the wavelength or the thickness of the glass is similar. The difference is that when the transmission is at the peak, the reflection is at the minimum because $R + T = 1$. For the glass with a thickness of 1.2 mm, the neighboring peaks of transmission separate for only 0.7 Å, which cannot be distinguished by the naked eye.

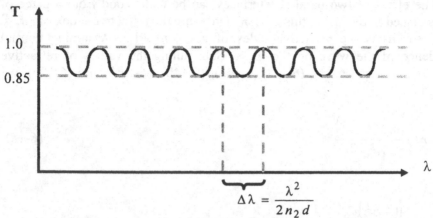

Figure 6-9. Variation of the light transmission through a piece of glass with the wavelength.

The periodic variation of reflection and transmission means that the two parallel interfaces cause interference of the wave propagating between them. The wave bouncing between the two interfaces is just like light oscillating between the mirrors of a laser cavity. For the wave propagating toward +x direction, if each round trip of this wave between the two interfaces has a phase difference of $2m\pi$ from the next round trip, constructive interference occurs and so transmission has its peak. Similarly, for the wave propagating toward the −x direction, when the round trips between the two interfaces have a phase difference of $2m\pi$, the reflection is at maximum due to the constructive interference. When the transmission has constructive interference, the reflection has destructive interference. The reason is that the corresponding waves propagate toward opposite directions, so they always have a path difference equal to the spacing of the two interfaces, i.e., the thickness of medium two.

2.3   Characteristics of Fabry-Perot Resonance

Most optical components are used in the situation with light transmission through the components. Thus we should look at the influence of Fabry-Perot resonance on the transmission of light through an optical component. The transmission of light intensity or power for an optical component is usually characterized by the transfer function T_{FPF}. From Eqs. (6-12), (6-13), (6-14a), (6-14b), (6-15a), and (6-15b), we obtain

$$T_{FPF} = \frac{(1-R)^2}{(1-R)^2 + 4R\sin^2[(\omega - \omega_0)d/n]} \tag{6-21}$$

where ω_0 is the wavelength that has maximum transmission. n and d are the refractive index and the thickness of the optical component respectively. R is the reflectivity of the individual interface between the optical component and its environment.

ω_0 is determined by the condition: $nd = $ m $(\lambda/2)$, where m is an integer. The spectral separation between the peaks of the transmission is called the free spectral range (FSR). The FSR in terms of wavelength $(\Delta\lambda)$ is given by Eq. (6-20). It can also be expressed in terms of frequency, $\Delta v = c/(2nd)$.

The peak of the transmission has a certain bandwidth. Its full width at half maximum, i.e., 3dB bandwidth, can be evaluated from Eq. (6-21) and is given by.

$$BW(3dB) = \frac{c}{2nd}\frac{1-R}{R^{1/2}} \tag{6-22}$$

2.4   Anti-Reflection Coating Using a Single Layer of Dielectric

As long as the refractive index of an optical component is not equal to one, its interface with the air always causes reflection of light. Such reflection is not desired in many cases. For example, the facet reflectivity of SOAs should be as small as possible, as we discussed in Chapter 4. In addition, almost every optical component has two interfaces with the air. If one of the interfaces is modified to exhibit some special function, the other interface still has reflection that could destroy or complicate the function. Furthermore, the Fabry-Perot resonance causes the reflection or transmission to rapidly oscillate with the wavelength. These effects are mostly undesirable. Anti-reflection (AR) coating is thus commonly used to reduce the reflectivity of the interface of optical components.

AR coating can be achieved using a single layer of dielectric or multiple layers of dielectric. The multi-layer coating is more complicated and will be discussed in the next section, while the condition for a single layer AR-coating can be easily derived. The condition is as follows. Assuming that the optical component has a refractive index n_3, a single layer of dielectric is coated on the optical component. The coated dielectric has the refractive index n_2 and outside the coating is air with refractive index $n_1 = 1$. Then the structure is the same as the one shown in Fig. 6-7. Therefore, the condition is that the reflection coefficient given in Eq. (6-11) is set to zero.

$$r = \frac{r_{12} + r_{23}e^{-2j\phi}}{1 + r_{12}r_{23}e^{-2j\phi}} = 0 \qquad (6\text{-}23)$$

It is equivalent to the requirement of $r_{12} + r_{23}e^{-2j\phi} = 0$. Because r_{12} and r_{23} are real, the only possibility to achieve this condition is that

$$e^{-2j\phi} = -1 \qquad (6\text{-}24)$$

$$r_{12} = r_{23} \qquad (6\text{-}25)$$

For normal incidence, Eqs. (6-24) and (6-25) lead to

$$d = \frac{\lambda}{n_2}(\frac{m}{2} + \frac{1}{4}) \qquad (6\text{-}26)$$

$$n_2 = \sqrt{n_1 n_3} \qquad (6\text{-}27)$$

where m is an integer. The variation of the reflectivity with the phase ϕ and the coating thickness d is shown in Fig.6-10. The minimum reflectivity is zero, occurring at the thickness phase $\phi = m\pi + \pi/2$ or the thickness given by Eq.(6-26).

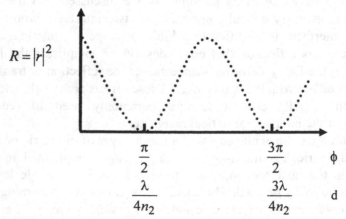

Figure 6-10. Variation of the reflectivity with the phase ϕ and the coating thickness d for the coating with the refractive index $n_2 = \sqrt{n_1 n_3}$.

3. TRANSMISSION AND REFLECTION FROM MULTI-LAYER COATINGS

Multi-layer coatings have much more complicated influences on the light wave than the single-layer coating. A straightforward analysis using the above approach is to calculate the continuity of the electric field and the magnetic field at the interfaces of those layers. However, such an approach is quite complicated because each layer will introduce two more undetermined coefficients and each interface will add two more equations. For many layers of coatings, to solve the total number of equations is very complicated. Therefore, a simpler approach is to evaluate the transmission and reflection from the matrix formulation [1] that will be introduced in this section.

3.1 Characteristic Matrix of a Single-Layer Coating

In this approach, the influence of each coating layer on the EM wave is represented by a matrix. Stacking many layers is then equivalent to cascading matrices. The matrix for a single layer is derived as follows.

Normal incidence on the coating is assumed and the surface of the coating is perpendicular to the z-axis, so the wave vector $\vec{k} = k\hat{z} = nk_0\hat{z}$, where n is the refractive index and $k_0 = (2\pi)/\lambda$. According to Faraday's law, if the wave is a plane wave, the electric field and the magnetic field are related by

$$-j\vec{k} \times \vec{E} = -jw\mu\vec{H} \tag{6-28}$$

From the above equation, we obtain $\dfrac{\omega\mu}{k_0}\vec{H} = n\hat{z} \times \vec{E}$, which can be rewritten as

$$Z_0\vec{H} = n\hat{z} \times \vec{E} \tag{6-29}$$

where $Z_0 = \sqrt{\dfrac{\mu}{\varepsilon_0}}$, called vacuum impedance.

The layer of coating is a thin film located between $z = Z_1$ and $z = Z_2$. Its refractive index is n_l. The refractive indices above and below this thin film are n_0 and n_s, respectively. The fields of the wave propagating toward the +z direction is represented with a superscript +, while those propagating toward the –z direction are represented with a superscript -. As shown in Fig. 6-11,

E^+ and H^+ are the electric field and the magnetic field of the wave propagating toward +z direction. On the other hand, the location right before the plane $z = Z_1$ is represented as Z_1^-. Similarly the location just after the plane $z = Z_1$ is represented as Z_1^+. Therefore, the electric field propagating from $z < Z_1$, arriving at the plane right before $z = Z_1$, yet meeting $z = Z_1$, is represented as $E^+(Z_1^-)$. Similarly, the electric field propagating from $z > Z_1$, arriving at the plane right after $z = Z_1$, already passing $z = Z_1$, is represented as $E^-(Z_1^-)$. Therefore, the total electric field right above the plane $z = Z_1$ is equal to $E^+(Z_1^-) + E^-(Z_1^-)$, while the total electric field right below the plane $z = Z_1$ is equal to $E^+(Z_1^+) + E^-(Z_1^+)$. The continuity of the tangential component of the electric field at the boundary $z = Z_1$ leads to

$$E^+(Z_1^-) + E^-(Z_1^-) = E^+(Z_1^+) + E^-(Z_1^+) = E(Z_1) \qquad (6\text{-}30)$$

From Eq. (6-29), we can obtain the magnetic field. Also, the tangential component of the magnetic field at the boundary $z = Z_1$ should be continuous. Therefore, we have

$$ZoH(Z_1^-) = [E^+(Z_1^-) - E^-(Z_1^-)]n_0$$

$$= [E^+(Z_1^+) - E^-(Z_1^+)]n_t = ZoH(Z_1^+) = ZoH(Z_1) \quad (6\text{-}31)$$

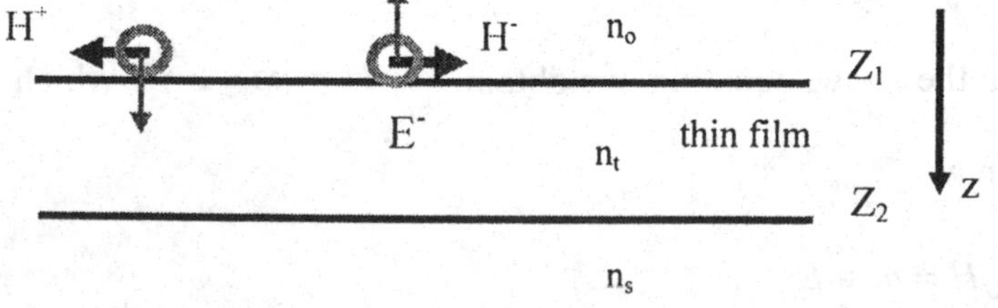

Figure 6-11. The electric field and the magnetic field right above the plane $z = Z_1$ for the wave propagating toward +z direction and –z direction.

Similarly, the continuity of the tangential component of the electric field and the magnetic field at the boundary $z = Z_2$ leads to the following equations.

$$E(Z_2) = E^+(Z_2^-) + E^-(Z_2^-) = E^+(Z_2^+) + E^-(Z_2^+) \qquad (6\text{-}32)$$

$$Z_0 H(Z_{2)} = [E^+(Z_2^-) - E^-(Z_2^-)]n_t$$

$$= [E^+(Z_2^+) - E^-(Z_2^+)]n_s \tag{6-33}$$

The electric fields $E^+(Z_1^+)$ and $E^+(Z_2^-)$ are in the thin film. The field $E^+(Z_2^-)$ is the field $E^+(Z_1^+)$ propagating from $z = Z_1^+$ to $z = Z_2^-$ in the thin film. Thus they are related by a phase delay.

$$E^+(Z_2^-) = E^+(Z_1^+)e^{-j\phi} \tag{6-34}$$

where

$$\phi - \frac{2\pi}{\lambda}n_t d \tag{6-35}$$

d is the thickness of the thin film.

Similarly, the field $E^-(Z_1^+)$ is the field $E^-(Z_2^-)$ propagating from $z = Z_2^-$ to $z = Z_1^+$ in the thin film, so $E^-\left(Z_1^+\right) = E^-\left(Z_2^-\right)e^{-j\phi}$, which is equivalent to

$$E^-(Z_2^-) = E^-(Z_1^+)e^{j\phi} \tag{6-36}$$

With some mathematical manipulation using Eqs.(6-30), (6-31), (6-32), (6-33), (6-34), and (6-36), we can obtain the following matrix equation.

$$\begin{bmatrix} E(Z_1) \\ Z_0 H(Z_1) \end{bmatrix} = \begin{bmatrix} \cos\phi & j\sin\phi/n_t \\ jn_t \sin\phi & \cos\phi \end{bmatrix} \begin{bmatrix} E(Z_2) \\ Z_0 H(Z_2) \end{bmatrix} \tag{6-37}$$

Eq.(6-37) means that the fields at the boundary $z = Z_2$ can be derived from the fields at the boundary $z = Z_1$ using the characteristic matrix of the film with the refractive index n_t. That is, the thin film can be represented by the matrix M_t. In addition, the matrix has nothing to do with the materials outside this thin film. In other words, the media n_0 and n_s have no influences on the matrix elements.

$$M_t = \begin{bmatrix} \cos\phi & j\sin\phi/n_t \\ jn_t \sin\phi & \cos\phi \end{bmatrix} \tag{6-38}$$

3.2   Characteristic Matrix for Multi-Layer Coatings

For double-layer coatings, each layer can be represented by a matrix like Eq. (6-38) with n_i replaced by the refractive index of this layer. As shown in Fig. 6-12, the first layer is between $z = Z_1$ and $z = Z_2$. The second layer is between $z = Z_2$ and $z = Z_3$. According to the above derivation, we should have

$$\begin{bmatrix} E(Z_1) \\ Z_0 H(Z_1) \end{bmatrix} = \begin{bmatrix} \cos\phi_1 & j\sin\phi_1 / n_1 \\ jn_1 \sin\phi_1 & \cos\phi_1 \end{bmatrix} \begin{bmatrix} E(Z_2) \\ Z_0 H(Z_2) \end{bmatrix} \tag{6-38}$$

and

$$\begin{bmatrix} E(Z_2) \\ Z_0 H(Z_2) \end{bmatrix} = \begin{bmatrix} \cos\phi_2 & j\sin\phi_2 / n_2 \\ jn_2 \sin\phi_2 & \cos\phi_2 \end{bmatrix} \begin{bmatrix} E(Z_3) \\ Z_0 H(Z_3) \end{bmatrix} \tag{6-39}$$

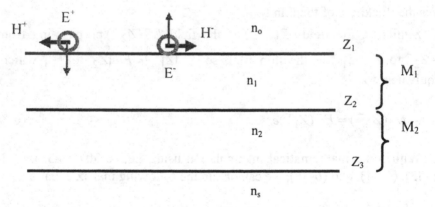

Figure 6-12. Double-layer coatings. Each layer can be represented by a matrix.

Combining Eqs. (6-38) and (6-39), we have

$$\begin{bmatrix} E(Z_1) \\ Z_0 H(Z_1) \end{bmatrix} = M_1 M_2 \begin{bmatrix} E(Z_3) \\ Z_0 H(Z_3) \end{bmatrix} \tag{6-40}$$

where

$$M_1 = \begin{bmatrix} \cos\phi_1 & j\sin\phi_1 / n_1 \\ jn_1 \sin\phi_1 & \cos\phi_1 \end{bmatrix} \tag{6-41a}$$

$$M_2 = \begin{bmatrix} \cos\phi_2 & j\sin\phi_2/n_2 \\ jn_2\sin\phi_2 & \cos\phi_2 \end{bmatrix} \qquad (6\text{-}41b)$$

For the same reason, if there are m layers of coatings, we will have the following result.

$$\begin{bmatrix} E(Z_1) \\ Z_0 H(Z_1) \end{bmatrix} = M_1 M_2 M_3 \cdots M_m \begin{bmatrix} E(Z_{m+1}) \\ Z_0 H(Z_{m+1}) \end{bmatrix} \qquad (6\text{-}42)$$

where

$$M_i = \begin{bmatrix} \cos\phi_i & j\sin\phi_i/n_i \\ jn_i\sin\phi_i & \cos\phi_i \end{bmatrix} \qquad i = 1, 2, 3, \ldots, m \qquad (6\text{-}43)$$

$$\phi_i = \frac{2\pi}{\lambda} n_i d_i$$

The multiplication of those matrices gives rise to a single matrix representing the multi-layer coatings.

$$M = M_1 M_2 M_3 \cdots M_m = \begin{bmatrix} M_{11} & M_{12} \\ M_{21} & M_{22} \end{bmatrix} \qquad (6\text{-}44)$$

The above matrix M is called the characteristic matrix of the multi-layer coatings. We have the fields at the top boundary $z = Z_1$ related to the fields at the bottom boundary $z = Z_{m+1}$ through the characteristic matrix of the multi-layer coatings.

$$\begin{bmatrix} E(Z_1) \\ Z_0 H(Z_1) \end{bmatrix} = M \begin{bmatrix} E(Z_{m+1}) \\ Z_0 H(Z_{m+1}) \end{bmatrix} \qquad (6\text{-}45)$$

3.3 Reflection and Transmission Coefficients of Multi-Layer Coatings

The fields of the incident wave are actually $E^+(Z_1^-)$ and $H^+(Z_1^-)$ instead of $E(Z_1)$ and $H(Z_1)$. Also, the reflected wave has the fields $E^-(Z_1^-)$ and $H^-(Z_1^-)$. On the other hand, the transmitted wave after the multi-layer coatings has the fields represented by $E^+(Z_{m+1}^+)$ and $H^+(Z_{m+1}^+)$, which are in the substrate with the refractive index n_s. If we consider the wave incidence only from the top, there will be no reflected wave in the substrate. Therefore, $E^-(Z_{m+1}^+) = H^-(Z_{m+1}^+) = 0$. Because we want to know the reflection and transmission coefficients, what we are interested in is the ratio of the field $E^-(Z_1^-)$ to the field $E^+(Z_1^-)$ and the ratio of the field $E^+(Z_{m+1}^+)$ to the field $E^+(Z_1^-)$. According to Eqs. (6-30), (6-31), (6-32), and (6-33), the fields $E^+(Z_1^-)$ and $E^-(Z_1^-)$ are related to $E(Z_1)$ and $H(Z_1)$. Also, the fields $E^+(Z_{m+1}^+)$ and $E^-(Z_{m+1}^+)$ are related to $E(Z_{m+1})$ and $H(Z_{m+1})$.

$$\begin{bmatrix} E^+\!\left(Z_1^-\right) \\ E^-\!\left(Z_1^-\right) \end{bmatrix} = \begin{bmatrix} \dfrac{1}{2} & \dfrac{1}{2n_0} \\ \dfrac{1}{2} & -\dfrac{1}{2n_0} \end{bmatrix} \begin{bmatrix} E(Z_1) \\ Z_0 H(Z_1) \end{bmatrix} \tag{6-46}$$

$$\begin{bmatrix} E(Z_{m+1}) \\ Z_0 H(Z_{m+1}) \end{bmatrix} = \begin{bmatrix} 1 & 1 \\ n_s & -n_s \end{bmatrix} \begin{bmatrix} E^+\!\left(Z_{m+1}^+\right) \\ E^-\!\left(Z_{m+1}^+\right) \end{bmatrix} \tag{6-47}$$

where $E^-(Z_{m+1}^+)=0$ because of no reflected wave after the last boundary z $= Z_{m+1}$.

Using Eqs. (6-45), (6-46) and (6-47), we obtain the relation between $E^-(Z_1^-)$ and $E^+(Z_1^-)$ as well as the relation between $E^+(Z_{m+1}^+)$ and $E^+(Z_1^-)$. Therefore, the reflection coefficient (r) and the transmission coefficient (t) are obtained.

$$r = \frac{E^-\!\left(Z_1^-\right)}{E^+\!\left(Z_1^-\right)} = \frac{n_0 M_{11} + n_0 n_s M_{12} - M_{21} - n_s M_{22}}{n_0 M_{11} + n_0 n_s M_{12} + M_{21} + n_s M_{22}} \tag{6-48}$$

$$t = \frac{E^+\left(Z_{m+1}^+\right)}{E^+\left(Z_1^-\right)} = \frac{2n_0}{n_0 M_{11} + n_0 n_s M_{12} + M_{21} + n_s M_{22}} \tag{6-49}$$

n_0 and n_s are the refractive indices above and below the multi-layer coating, respectively. The above ratios are for the amplitude of fields. The reflectivity and the transmission of power or intensity are given by the square of the above quantities.

$$R = |r|^2 \tag{6-50}$$

$$T = \frac{n_s}{n_0}|t|^2 \tag{6-51}$$

3.4   Inclined Incidence

The above discussion is for normal incidence of the wave. If the wave is not normally incident on the surface, the characteristic matrix of each layer changes. In this case, the TE wave and the TM wave behave differently, although the derivation procedure for each will be similar. The results are similar to the previous derivations except that the refractive indices and the phase delay in Eq. (6-43) changed to the following values.

For TM wave: $n_i \rightarrow n_{pi} = \dfrac{n_i}{\cos\theta_i}$ $\tag{6-52}$

For TE wave: $n_i \rightarrow n_{si} = n_i \cos\theta_i$ $\tag{6-53}$

$$\phi_i = \frac{2\pi}{\lambda} n_i d_i \rightarrow \phi_i = \frac{2\pi}{\lambda} n_i d_i \cos\theta_i \tag{6-54}$$

where θ_i is the angle between the wave vector and the normal of the boundary in the layer i. This angle satisfies Snell's law.

$$n_1 \sin\theta_1 = n_2 \sin\theta_2 = \cdots = n_i \sin\theta_i = \cdots = n_m \sin\theta_m \tag{6-55}$$

3.5   Practical Design of Multi-Layer Coatings for Desired Purposes

According to Eqs. (6-43) and (6-44), the matrix elements are functions of the refractive index and the phase delay in each layer, while the phase delay is a function of the layer thickness and the wavelength. Therefore, as long as we know the refractive index and the thickness of each layer as well as the wavelength, we are able to calculate the reflection and transmission coefficients using Eqs (6-48) and (6-49). If the incidence of the wave is not normal, we only need to replace those parameters with the formula given in Eqs. (6-52), (6-53) and (6-54). This calculation is simple. The process of calculating the reflection and the transmission from the known layer structure is called "analysis".

In practical applications, we usually have the desired function first. Then we design the layer structure of the multi-layer coatings according to the desired function. For example, we want to have a beam splitter working at the wavelength between 1.3 μm and 1.55 μm with the reflection and transmission of the optical power both at 50 %. We then have to determine the combination of the refractive index, the thickness for each layer, and the number of layers. This is not an easy task.

To design a layer structure of coatings according to a known function is called "synthesis". Synthesis is much more difficult than analysis. Furthermore, there is usually more than one solution that can achieve the desired function. To obtain a good structure of multi-layer coatings for certain purposes requires lots of experience with coating design. [2] Fortunately, quite a few software packages are available for the design, making it possible for less experienced engineers to carry out the work.

In the design of coatings, other conditions also have to be considered. First, the refractive index cannot be randomly selected. It has to be the value of a real material. Second, the selected layer of coating has to adhere well to the substrate or the layer on which it is coated. Third, those layers of coatings should sustain the variation of temperature and humidity required in real applications. Forth, there should be deposition or growth equipment to apply the selected materials. In practical fabrication of the coated layers, very few types of materials are used. Typically only two or three types of materials are used for a given purpose, while the total number of layers is numerous.

The derivation above is based on plane-wave approximation. In optical communications, a laser beam is usually used. Because the laser beam is a Gaussian beam, its reflection and transmission from the multi-layer coatings deviates slightly from that of a plane wave. Fortunately, the deviation is mostly negligible. Also, the coating may be applied to a surface that is not a plane. For example, a lens with a spherical surface is coated for the reduction of reflectivity. As long as the radius of curvature is large the above derivation is still useful for lens coating calculation.

3.6 Reflection from Bragg Gratings

A Bragg grating is formed by stacking pairs of half-wavelength or quarter-wavelength layers that alternate high refractive index and low refractive index. Pairs of half-wavelength layers with alternating high refractive index and low refractive index are used in DFB LDs and DBR LDs. Pairs of quarter-wavelength layers with alternating high refractive index and low refractive index are used for VCSELs. The Bragg grating has many layers of different refractive indices that behave like the multi-layer coatings described above, so it can be analyzed using the approach of the characteristic matrix.

For pairs of quarter-wavelength layers used for VCSELs, we have the characteristic matrix for the layer of low refractive index as follows.

$$M_l = \begin{bmatrix} 0 & j/n_l \\ jn_l & 0 \end{bmatrix} \qquad (6\text{-}56)$$

It is evaluated using $\phi = \dfrac{2\pi}{\lambda} nd$ for $d = \dfrac{\lambda}{4n}$ in Eq. (6-43). Similarly, the characteristic matrix for the layer of high refractive index is given by

$$M_h = \begin{bmatrix} 0 & j/n_h \\ jn_h & 0 \end{bmatrix} \qquad (6\text{-}57)$$

For one pair of the above quarter-wavelength layers, the characteristic matrix is the product of the matrices in Eqs. (6-56) and (6-57).

$$M_{pair} = \begin{bmatrix} (-\dfrac{n_h}{n_l}) & 0 \\ 0 & (-\dfrac{n_l}{n_h}) \end{bmatrix} \qquad (6\text{-}58)$$

For N pairs of quarter-wavelength layers with alternating high refractive index and low refractive index, the characteristic matrix is then the product of the above matrix with itself N times.

$$M_{N-pairs} = \begin{bmatrix} (-\dfrac{n_h}{n_l})^N & 0 \\[2ex] 0 & (-\dfrac{n_l}{n_h})^N \end{bmatrix} \tag{6-58}$$

Substituting the matrix elements in Eq. (6-58) into Eq. (6-48), we obtain the reflection coefficient r as follows.

$$r = \frac{n_0(-\dfrac{n_h}{n_l})^N - n_s(-\dfrac{n_l}{n_h})^N}{n_0(-\dfrac{n_h}{n_l})^N + n_s(-\dfrac{n_l}{n_h})^N} = \frac{1 - \dfrac{n_s}{n_0}(\dfrac{n_l}{n_h})^{2N}}{1 + \dfrac{n_s}{n_0}(\dfrac{n_l}{n_h})^{2N}} \tag{6-59}$$

As we described before, n_0 is the refractive index of the initial medium from which the wave is incident. n_s is the refractive index of the last medium that is at the far side of the stacking pairs. The reflectivity is the square of the above reflection coefficient.

The above calculation is for the wavelength that is equal to four times the layer thickness. For another wavelength, similar calculation can be done, but with more complicated procedure because the phase ϕ is not equal to $\pi/2$. In fact, the stacking pairs of quarter-wavelength layers with alternating high refractive index and low refractive index perform like a band pass filter.[3] The variation of the reflectivity with the wavelength is shown in Fig. 6-13.

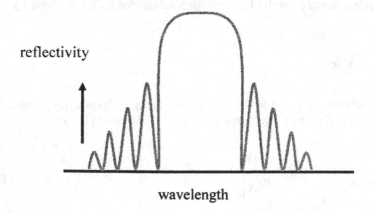

Figure 6-13. The variation of the reflectivity with the wavelength for stacking pairs of quarter-wavelength layers with alternating high refractive index and low refractive index.

4.   DIFFRACTION GRATING

A diffraction grating has a surface with periodic variation of topography. The period is close to the wavelength of light. A grating with a corrugated surface is schematically shown in Fig. 6-14. As light is incident on one period of the surface variation, it is reflected toward many directions because the surface is not a plane. In Fig. 6-14, the orange arrow lines indicate the incident light, while the pink, blue, and red arrow lines show a few examples of the reflected light. Light shining on two periods of the corrugated surface shown in the figure.

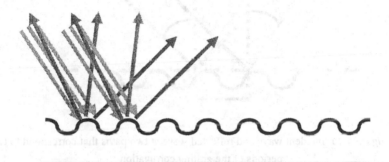

Figure 6-14. A schematic of grating surface. Light shining on two periods of the corrugated surface is shown. The orange arrow lines indicate the incident light, while the pink, blue, and red arrow lines show a few examples of the reflected light.

4.1   Principle of Grating

The path length for the light reflected at different periods is not the same. Fig. 6-15 shows that light of path one is reflected at the first grating, while light of path two is reflected at the second grating. The blue dashed line represents an imagined surface that connects the same height of corrugation in each period. It is a flat plane. The incident angle θ_i is defined as the angle between the normal of this flat plane and the incident direction of light. The diffraction angle θ_r is defined as the angle between the normal of the same plane and the direction of reflected light. θ_i is not necessarily equal to θ_r because the local surface is not always in parallel with the imagined plane shown by the blue dashed line.

When the light in path one and in path two is at the same wave front, they are in phase. As shown in Fig. 6-16 (a), when the wave front of the light in path one has arrived at the first grating, the same wave front in the path two has not arrived at the second grating. This wave front is indicated by the orange dashed line. The same wave front in path two has to travel a further distance to reach the grating. The distance is equal to $\Lambda\sin\theta_i$, as shown in Fig. 6-16 (a). On the other hand, the wave front of the reflected light in path one leaves the grating earlier than the same wave front in path two for a distance $\Lambda\sin\theta_r$, as shown in Fig. 6-16 (a). The wave front of the reflected wave is indicated by the red dashed line.

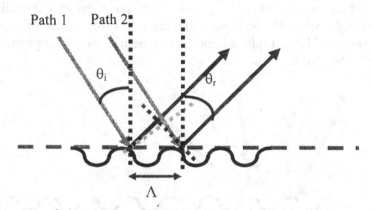

Figure 6-15. Incident wave and reflected wave at two paths that correspond to two periods of the grating corrugation.

Therefore, Fig. 6-16 (a) shows that, from the wave front (orange dashed line) to the grating, path two is longer than path one for a distance of $\Lambda\sin\theta_i$ for the incident wave. For the reflected wave, Fig. 6-16 (b) shows that path one is longer than path two for a distance of $\Lambda\sin\theta_r$ from the grating to the wave front (red dashed line). As a result, the path difference Δ between path one and path two is given by

$$\Delta = \Lambda\sin\theta_r - \Lambda\sin\theta_i \qquad\qquad (6\text{-}60)$$

As the light is reflected by many periods of corrugation, the reflected light is only significant when the path difference is a multiple of the wavelength due to the constructive interference. Otherwise, the reflect light from different periods of corrugation will cancel one another, leading to very weak reflected light. Therefore, we have the principle of grating as follows

$$\sin\theta_r = \sin\theta_i + m\frac{\lambda}{\Lambda} \qquad\qquad (6\text{-}61)$$

where m is an integer. θ_r is the angle that significant intensity of diffraction light can be measured.

(a)

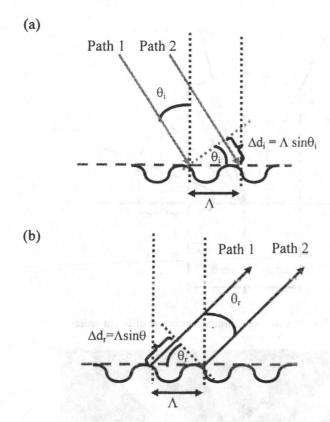

(b)

Figure 6-16. The path difference for (a) incident wave and (b) reflected wave.

4.2 Effect of Grating Diffraction

For m=0, the diffraction angle is the same as the incident angle, so this diffraction light is like the reflection from a normal mirror. This light is called 0 order. For m = 1, the diffraction light at this angle is called +1 order. For m = -1, the diffraction light at this angle is called -1 order. For other values of m, those orders are named similarly. The directions of those orders are shown in Fig. 6-17.

According to Eq. (6-61), the diffraction angle θ_r depends on the wavelength for m ≠ 0. Therefore, the grating separates light of different wavelengths to propagate toward different directions. The larger the wavelength λ, the larger the diffraction angle. Fig. 6-18 shows the diffraction of different colors if the incident light contains the three colors: blue, green, and red. As long as m ≠ 0, the red light is always farther away from the 0 order than the blue light. On the other hand, the smaller period of grating gives rise to a larger variation of the diffraction angle with the wavelength.

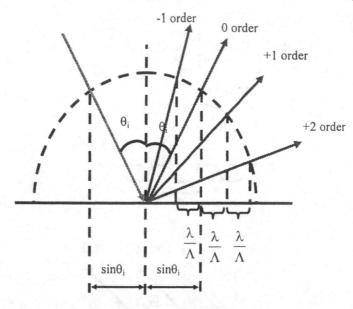

Figure 6-17. The directions of diffraction beams at different orders.

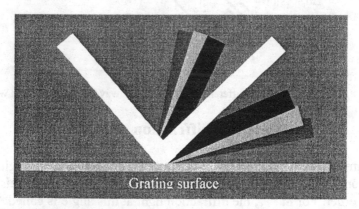

Figure 6-18. Diffraction of different colors of light by the grating for the incident light containing blue, green, and red colors.

Although significant light is expected at the angles given by Eq. (6-61), the intensity of light at different orders is not the same. The distribution of the intensity over those orders depends on the shape of the corrugated surface in each period. [4] The calculation of the intensity distribution is very complicated and will not be given here. One common design is where the strongest intensity is at either the +1 order or the −1 order. Nearly 90 % of the power of the incident light can be possibly obtained at either the +1 order or the −1 order with proper design of the grating topography.

4.3   Spectral Resolution

For the diffraction light at a particular order, the power of light is not completely concentrated at the diffraction angle given by Eq. (6-61). In fact, the power is distributed over a range of angles. On the other hand, although the light of a particular wavelength has a diffraction angle given by Eq.(6-61), some portion of the light's power at other wavelengths can also be diffracted to this angle. Therefore, as we measure the light at this diffraction angle, we will obtain the power of light distributed over a spectral range, as illustrated in Fig. 6-19.

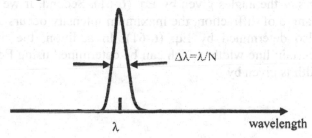

Figure 6-19. Distribution of light power over the wavelength at the diffraction angle.

The full width at half maximum (FWHM) of the spectrum can be calculated by using detailed analysis of the overall interference effect of light reflected from every period of corrugation. The path difference for the light reflected from two neighboring periods is given by Eq. (6-60). This path difference will cause a phase difference

$$\Delta\phi = 2\pi \frac{\Delta}{\lambda} \tag{6-62}$$

Assuming the light reflected from the first period has a phase of ϕ_0, the light reflected from the second period, the third period, . . ., etc. will have a phase of $\phi_0 + \Delta\phi$, $\phi_0 + 2\Delta\phi$, . . .,etc. As we measure at the direction of the diffraction angle θ_r, the light reflected from different periods of the grating add up. Thus we have the total field proportional to the following term.

$$A(N) = \sum_{m=0}^{N-1} e^{-j(\phi_0 + m\Delta\phi)} \tag{6-63}$$

The intensity of light is proportional to the square of the absolute value of $A(N)$.

$$I \propto \left| A(N) \right|^2 = \frac{\sin^2(N\Delta\phi/2)}{\sin^2(\Delta\phi/2)} \tag{6-64}$$

$\dfrac{\sin^2(N\Delta\phi/2)}{\sin^2(\Delta\phi/2)}$ has maxima at $\Delta\phi = 2m\pi$, where m is an integer. Using $\Delta\phi$ in Eq. (6-62) and Δ in Eq. (6-60), we obtain the same principle of grating given in Eq. (6-61). We can look at the maxima from two view points. First, if the incident light has only one wavelength, the maximum intensity of diffraction occurs at the angles given by Eq. (6-61). Second, if we measure light at a fixed angle of diffraction, the maximum intensity occurs at certain wavelengths, also determined by Eq. (6-61). In addition, the maximum intensity has a certain line-width, which can be determined using Eq. (6-64). The FWHM width is given by

$$\Delta\lambda = \frac{\lambda}{N} \tag{6-65}$$

where N is the total number of grating periods that the light beam covers. The peak power is at the wavelength that is related to the diffraction angle given by Eq.(6-61).

5.   WAVEGUIDE COUPLING

A waveguide like an optical fiber is used to make a light wave propagate along the desired route. The waveguide is usually formed with the structure where a dielectric of a high refractive index is surrounded by another dielectric of a low refractive index. The center part of the dielectric with the high refractive index is called the core, while the surrounding dielectric of the low refractive index is called the cladding. The light power of the guiding mode is mostly confined within the core. However, the tail of the mode profile of the guiding mode usually extends to the cladding, as shown in Fig. 6-20.

When such a waveguide is placed very close to another waveguide, the tail of its mode profile might extend to another waveguide. As a result, the portion that exists in another waveguide may grow in its intensity. The power of light in the original waveguide may then couple to the guiding mode in another waveguide. The situation is illustrated in Fig. 6-21. The top waveguide has a core with the refractive index n_1, while the bottom waveguide has a core with the refractive index n_2. The cores are surrounded by cladding regions with a refractive index n, which is smaller than n_1 and n_2. The guiding mode in the top waveguide gradually couples to the bottom waveguide and then becomes guiding in the bottom one.

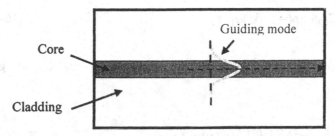

Figure 6-20. The guiding mode of light wave propagating in a waveguide. The guiding mode extends to the cladding region.

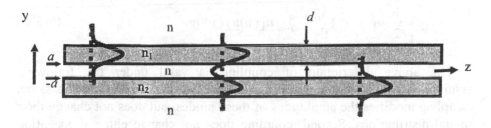

Figure 6-21. Coupling of the guiding mode from the top waveguide to the bottom waveguide.

5.1 Coupling Equations

The coupling of the guiding mode from one waveguide to another can be analyzed as follows. [5] The structures of waveguides shown in Fig.6-20 are used for the analysis. The waveguides are aligned along the z-axis. The core has the width of d. The top waveguide has its core extending from y = a to $a+d$. The bottom waveguide has its core extending from y = -a to -a-d. The field distribution of the guiding mode across the core width in the top waveguide and the bottom waveguide is represented by $u_1(y)$ and $u_2(y)$, respectively. When the guiding modes propagate along the z-axis, the mode profile remains the same, while its amplitude varies. The amplitudes are represented by $a_1(z)$ and $a_2(z)$, respectively, for the modes in the top waveguide and the bottom waveguide. Therefore, the guiding modes in the top waveguide and the bottom waveguide are represented by $a_1(z)u_1(y)e^{-j\beta_1 z}$ and $a_2(z)u_2(y)e^{-j\beta_2 z}$, respectively, where β_1 and β_2 are the propagation constants.

As the guiding modes propagate along the z-axis, coupling occurs. Then the amplitudes $a_1(z)$ and $a_2(z)$ vary according to the following equations.

$$\frac{da_1}{dz} = -jC_{21}\exp(j\Delta\beta z)a_2(z) \tag{6-66a}$$

$$\frac{da_2}{dz} = -jC_{12}\exp(-j\Delta\beta z)a_1(z) \tag{6-66b}$$

where $\Delta\beta = \beta_1 - \beta_2$ and the coupling coefficients C_{12} and C_{21} are given by

$$C_{21} = \frac{1}{2}(n_2^2 - n^2)\frac{k_o^2}{\beta_1}\int_a^{a+d} u_1(y)u_2(y)dy \tag{6-67a}$$

$$C_{12} = \frac{1}{2}(n_1^2 - n^2)\frac{k_o^2}{\beta_2}\int_{-a-d}^{-a} u_1(y)u_2(y)dy \tag{6-67b}$$

The above description of coupling is valid under the following assumptions. First, the mode profile in each waveguide remains the same. Coupling modifies the amplitudes of these modes, but does not change their spatial distributions. Second, coupling does not change either propagation constant. Third, the coupling coefficients are given by the proportion of modal overlap, as shown in Eqs. (6-67a) and (6-67b).

5.2   Solutions of Coupling Equations

Eqs. (6-66a) and (6-66b) can be easily solved. Assume that the guiding mode is only in the top waveguide for z = 0, so $a_2(z = 0) = 0$. Then the solutions are.

$$a_1(z) = a_1(0)\exp(\frac{j\Delta\beta z}{2})(\cos\gamma z - j\frac{\Delta\beta}{2\gamma}\sin\gamma z) \tag{6-68a}$$

$$a_2(z) = a_1(0)\frac{C_{12}}{j\gamma}\exp(-\frac{j\Delta\beta z}{2})\sin\gamma z \tag{6-68b}$$

where $\gamma^2 = (\frac{\Delta\beta}{2})^2 + C^2$ and $C = (C_{12}C_{21})^{1/2}$.

The optical power $P_1(z)$ in the top waveguide and $P_2(z)$ in the bottom waveguide are proportional to $|a_1(z)|^2$ and $|a_2(z)|^2$, respectively, so

$$P_1(z) = P_1(0)[\cos^2 \gamma z + (\frac{\Delta\beta}{2\gamma})^2 \sin^2 \gamma z] \tag{6-69a}$$

$$P_2(z) = P_1(0)\frac{|C_{12}|^2}{\gamma^2}\sin^2 \gamma z \tag{6-69b}$$

When there is no loss in the waveguide, we see that $P_1(z) + P_2(z) = P_1(0)$. The total power is conserved. The ratio of power coupling from one waveguide to another depends on the distance (z), the coupling coefficients and the difference of propagation constants of the guiding modes in the two waveguides. An example of the power variation with the propagation distance during the coupling procedure is shown in Fig. 6-22. At the beginning, the power in the top waveguide decreases and reaches a minimum. In the meantime, the power in the bottom waveguide increases and reaches a maximum. Over this section, the guiding mode couples from the top waveguide to the bottom waveguide. Thereafter, the power of the guiding mode in the top waveguide increases, while the power of guiding mode in the other waveguide decreases. The power variation is a periodic function of the propagation distance with a period of π/γ.

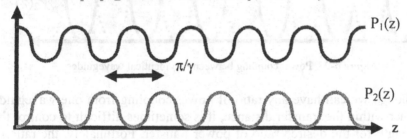

Figure 6-22. An example of the power coupling between two waveguides.

When the two waveguides are identical, we have $\beta_1 = \beta_2$. Then $C = \gamma$. This situation is called phase matched. The power equations become

$$P_1(z) = P_1(0)\cos^2 \gamma z \tag{6-70a}$$

$$P_2(z) = P_1(0)\sin^2 \gamma z \tag{6-70b}$$

The power variation with the propagation distance for this coupling procedure is shown in Fig. 6-23. The transfer of power from one waveguide to another one can be 100 %. The power in the top waveguide decreases at the beginning and then reduces to zero. The complete transfer of power occurs at $z = \dfrac{\pi}{2C} \equiv L_0$. L_0 is called transfer distance.

5.3   Control of Coupling Ratio

In practical applications, we do not require the coupling distance to exceed the transfer distance because we can have any ratio of power coupling from one waveguide to another within this range. For example, at a distance of $L_0/2$, we have 50 % of the power transferred from one waveguide to another, so the device works as a 3-dB coupler. If the two closely spaced waveguides separate with a larger space after 50 % of the power transfer, the two waveguides will have equal power afterwards. That is, the initial power in one waveguide is split equally between the two waveguides.

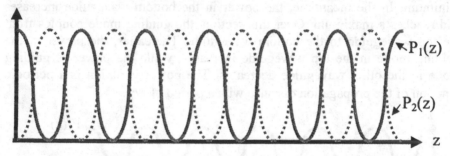

Figure 6-23. Power coupling between two identical waveguides.

Although we can have any ratio of power coupling from one waveguide to another within the transfer distance, it is sometimes difficult to control the exact length for the desired ratio of power transfer. Fortunately, the ratio of power transfer can be controlled using a small phase mismatch $\Delta\beta$. With a fixed length of $L_0 = \dfrac{\pi}{2C}$ and using $\gamma^2 = (\dfrac{\Delta\beta}{2})^2 + C^2$ and Eq. (6-69b), we obtain the ratio of power transfer as a function of the phase mismatch $\Delta\beta$.

$$T = \frac{P_2(L_0)}{P_1(0)} = (\frac{\pi}{2})^2 \sin c^2 \left\{ \frac{1}{2}[1 + (\frac{\Delta\beta L_0}{\pi})^2]^{1/2} \right\} \tag{6-71}$$

where $\sin c(x) = \dfrac{\sin(\pi x)}{\pi x}$. The variation of this ratio with the phase mismatch parameters, $\Delta\beta L_0$, is plotted in Fig. 6-24. The power-transfer ratio is 100% for $\Delta\beta = 0$ and decreases as the phase mismatch increase. It becomes zero when $\Delta\beta L_0 = \sqrt{3}\pi$. Therefore, the ratio of power transfer can be controlled to be any value between zero and 100 % by properly choosing $\Delta\beta L_0$ between 0 and $\sqrt{3}\pi$.

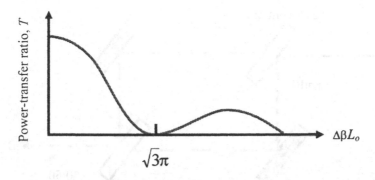

Figure 6-24. Variation of the power-transfer ratio with the phase mismatch parameters, $\Delta\beta L_0$,

6. MACH-ZEHNDER INTERFERENCE

The Mach-Zehnder interferometer is a common device used in several kinds of optical components for optical communication. A simple concept for the Mach-Zehnder interferometer is to separate a light beam into two different paths and then recombine them. Interference occurs due to the phase difference between the two paths when the two split beams recombine. The concept of Mach-Zehnder interference can be explained using the configuration shown in Fig. 6-25. The input light beam is separated to path one and path two using a 50/50 beam splitter, which means that the intensity of either reflection or transmission is 50% of the incident light intensity. Thus the two split light beams have equal intensity. Both light beams are reflected by a mirror and then meet again at the second 50/50 beam splitter. The second 50/50 beam splitter should again separate the beam in each path equally to the output A and the output B. If only one beam is incident on the second beam splitter, this is true. However, when we have both beams simultaneously incident on the second beam splitter, the situation is no longer straightforward. The output A and the output B do not equally share the light intensity. The intensities of light at the output A and the output B depend on the phase difference between the two paths. If the light traveling in path one and the light traveling in path two have a phase difference of $\Delta\phi$ when they arrive at the second beam splitter, the intensities at the two outputs will be

$$I_A \propto 1+\cos\Delta\phi \qquad (6\text{-}72a)$$

$$I_B \propto 1+\cos(\Delta\phi+\pi) \qquad\qquad (6\text{-}72\text{b})$$

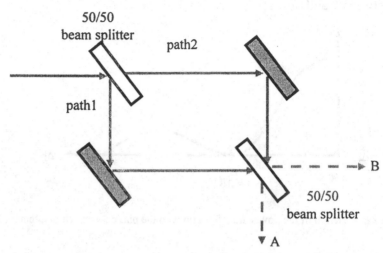

Figure 6-25. A configuration of optical setup to demonstrate Mach-Zehnder interference.

Therefore, the intensities of output A and output B are a function of the phase difference $\Delta\phi$, as shown in Fig. 6-26. From Fig. 6-26, we see that the intensities at output A and output B can be controlled by the phase difference. For example, if we choose a phase difference of 0, the intensity is concentrated at output A. If the phase difference is π, the intensity is switched to output B. If the phase difference is $\pi/2$, we have equal intensities at the two outputs.

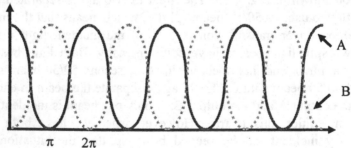

Figure 6-26. Variation of the intensities of the output A and the output B with the phase difference $\Delta\phi$.

Using this concept, a waveguide-guide type Mach-Zehnder interferometer can be constructed. Fig. 6-27 shows an example. The 3dB coupler is a waveguide coupler as described in the previous section. 3dB means that the power ratio of 50%, when comparing the output power to the input power. The power from the input is split by the 3dB coupler equally into the two waveguides in the center part, indicated as path one and path two. Afterwards, they are combined using the second 3dB coupler. The power at output A and output B then depend on the phase difference of the light passing through the two paths. Choosing proper lengths of the waveguides in path one and path two, we can intentionally create a phase difference to achieve a certain ratio of the power between the output A and the output B. On the other hand, if the phase of light in either path or both paths is controlled by an external voltage, the power ratio between the output A and the output B becomes tunable.

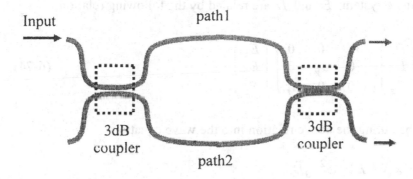

Figure 6-27. An example of the waveguide-guide type Mach-Zehnder interferometer.

7. BIREFRINGENCE

In Chapter 1, we briefly discussed the origin of birefringence. When an electric field is applied to materials in which the atomic structure is not symmetric, the positive and negative charges do not separate along the direction of applied electrical field due to different bonding forces among the different directions. As a result, the polarization $\vec{P}$, the electrical field $\vec{E}$, and the field $\vec{D}$ are not necessarily parallel. Such materials are called anisotropic materials. In these materials, the dielectric constant ε depends on the direction of electric field, so the refractive index ($=\sqrt{\mu\varepsilon}/\sqrt{\mu_0\varepsilon_0}$) is not a constant. This phenomenon leads to birefringence. The detailed analysis is given in the following.[6]

7.1 Dispersion Relation between *k* and ω

Because $\vec{E}$ and $\vec{D}$ are generally not parallel in anisotropic materials, they are related by a tensor of dielectric constant. The relation is as follows.

$$\begin{pmatrix} D_x \\ D_y \\ D_z \end{pmatrix} = \begin{pmatrix} \varepsilon_{xx} & \varepsilon_{xy} & \varepsilon_{xz} \\ \varepsilon_{yx} & \varepsilon_{yy} & \varepsilon_{yz} \\ \varepsilon_{zx} & \varepsilon_{zy} & \varepsilon_{zz} \end{pmatrix} \begin{pmatrix} E_x \\ E_y \\ E_z \end{pmatrix} \tag{6-73}$$

It is possible to transform the coordinate system for this relation to another so that the tensor of dielectric constant ε is diagonal in the new system. This system is called the principal coordinate system, which is determined by the atomic structure of the materials. In the principal coordinate system, $\bar{E}$ and $\bar{D}$ are related by the following relation

$$\begin{pmatrix} D_x \\ D_y \\ D_z \end{pmatrix} = \begin{pmatrix} \varepsilon_x & 0 & 0 \\ 0 & \varepsilon_y & 0 \\ 0 & 0 & \varepsilon_z \end{pmatrix} \begin{pmatrix} E_x \\ E_y \\ E_z \end{pmatrix} \tag{6-74}$$

Substituting the above relation into the wave equation

$$\nabla \times \left(\nabla \times \bar{E} \right) = \omega^2 u_0 \bar{D}$$

we obtain the following dispersion relation for the plane wave.

$$\begin{pmatrix} \omega^2 \mu \varepsilon_x - k_y^2 - k_z^2 & k_x k_y & k_x k_z \\ k_y k_x & \omega^2 \mu \varepsilon_y - k_x^2 - k_z^2 & k_y k_z \\ k_z k_x & k_z k_y & \omega^2 \mu \varepsilon_z - k_x^2 - k_y^2 \end{pmatrix} \begin{pmatrix} E_x \\ E_y \\ E_z \end{pmatrix} = 0 \tag{6-75}$$

To have nontrivial solutions of the field $\bar{E}$, the determinant of the above matrix must equal zero.

$$\begin{vmatrix} \omega^2 \mu \varepsilon_x - k_y^2 - k_z^2 & k_x k_y & k_x k_z \\ k_y k_x & \omega^2 \mu \varepsilon_y - k_x^2 - k_z^2 & k_y k_z \\ k_z k_x & k_z k_y & \omega^2 \mu \varepsilon_z - k_x^2 - k_y^2 \end{vmatrix} = 0 \tag{6-76}$$

Solve Eq. (6-76) and obtain the relation between k and ω. Then the electric filed can be solved.

$$k = k(\omega) \tag{6-77}$$

$$\begin{pmatrix} E_x \\ E_y \\ E_z \end{pmatrix} \propto \begin{pmatrix} \dfrac{k_x}{k^2 - \omega^2 \mu \varepsilon_x} \\ \dfrac{k_y}{k^2 - \omega^2 \mu \varepsilon_y} \\ \dfrac{k_z}{k^2 - \omega^2 \mu \varepsilon_z} \end{pmatrix}$$

(6-78)

The explicit form of Eq.(6-77) depends on the isotropic property of the materials.

7.2   Dispersion Relation for Isotropic Materials

For isotropic materials, ε is the same for all directions, so $\varepsilon_x = \varepsilon_y = \varepsilon_z$. Substituting them into Eq. (6-76), we have the simple relation between k and ω as follows.

$$\frac{k^2}{n^2} - \frac{\omega^2}{c^2} = 0$$

(6-79)

where

$$n^2 = \frac{\varepsilon_x}{\varepsilon_0} = \frac{\varepsilon_y}{\varepsilon_0} = \frac{\varepsilon_z}{\varepsilon_0}$$

(6-80)

Eq. (6-79) reduces to the simple relation $k = n\omega/c$, which is well known in fundamental EM theory.

7.3   Dispersion Relation for Uniaxial Materials

For anisotropic materials, the situation is divided into two different cases. The first case is that two components of the dielectric constant are the same, while the third one is different. This case is called uniaxial. Usually, the two equal constants are designated ε_x and ε_y. The different constant is designated ε_z. Thus $\varepsilon_x = \varepsilon_y \neq \varepsilon_z$. Substituting them into Eq. (6-76), we have the relation between k and ω as follows.

$$\left(\frac{k_x^2 + k_y^2}{n_e^2} + \frac{k_z^2}{n_0^2} - \frac{\omega^2}{c^2}\right)\left(\frac{k^2}{n_0^2} - \frac{\omega^2}{c^2}\right) = 0 \tag{6-81}$$

where $n_0^2 = \dfrac{\varepsilon_x}{\varepsilon_0} = \dfrac{\varepsilon_y}{\varepsilon_0}$, $n_e^2 = \dfrac{\varepsilon_z}{\varepsilon_0}$.

Eq. (6-81) means that there are two possible solutions of k for a given frequency ω. The first one is determined by the equation

$$\frac{k^2}{n_0^2} - \frac{\omega^2}{c^2} = 0 \tag{6-82a}$$

The second one is determined by

$$\frac{k_x^2 + k_y^2}{n_e^2} + \frac{k_z^2}{n_0^2} - \frac{\omega^2}{c^2} = 0 \tag{6-82b}$$

For a given value of ω, the solutions of k from these two equations actually form two surfaces, called normal surfaces. The surface determined by Eq.(6-82a) is a sphere and the one determined by Eq.(6-82b) is an ellipsoid. They intersect at the k_z-axis. Therefore, there are two k values for a given ω except at the z-direction. Because the refractive index is determined by k/k_0, where k_0 is the wave vector in the free space, we have two different refractive indices for a wave propagating in the anisotropic materials.

Solving Eq. (6-82a), we have $k = n_0\omega/c$. This value is independent of the direction of wave propagation. From this k value, we obtain the refractive index equal to n_0. The wave that experiences this refractive index is called the ordinary wave because its refractive index is like the usual wave in the isotropic material.

The k value determined by Eq.(6-82b) depends on the direction of the wave vector $\bar{k}$, so the refractive index also depends on the direction of the wave vector $\bar{k}$. For example, for the wave vector $\bar{k}$ that is on the y-z plane, but is at an angle θ from the z-axis, we have $k_x = 0$, $k_y = n_e(\theta)\dfrac{\omega}{c}\sin\theta$, and

$k_z = n_e(\theta)\dfrac{\omega}{c}\cos\theta$. Substituting them into Eq. (6-82b), we obtain

$$\frac{1}{n_e^2(\theta)} = \frac{\cos^2\theta}{n_0^2} + \frac{\sin^2\theta}{n_e^2} \tag{6-83}$$

For $\theta = 0$, i.e., the wave vector $\bar{k}$ along the z-axis, the refractive index is equal to n_0. If $\theta = \pi/2$, i.e., wave vector $\bar{k}$ along the y-axis, the refractive index is equal to n_e. For the angle between 0 and $\pi/2$, the refractive index $n_e(\theta)$ is between n_0 and n_e. Because the behavior of the refractive index is different from the usual situation of isotropic materials, the wave that experiences such a refractive index is called extraordinary wave. The electric field of the ordinary wave and the electric field of the extraordinary wave are perpendicular to one another.

7.4   Dispersion Relation for Biaxial Materials

The second case of the anisotropic materials is that the ε_x, ε_y, and ε_z are all different. This case is called biaxial. The solutions for $k = k(\omega)$ are quite complicated. For a given solution of ω, the solutions of k form two shells that intersect at four points. Therefore, we have two k values again for a given ω except at the four intersection points, so there are two different refractive indices for a wave propagating in the anisotropic materials.

7.5   Index Ellipsoid

In general, the index of an anisotropic material can be described by the index ellipsoid in the principal coordinate system

$$\frac{x^2}{n_x^2} + \frac{y^2}{n_y^2} + \frac{z^2}{n_z^2} = 1 \tag{6-84}$$

where

$$n_x^2 = \frac{\varepsilon_x}{\varepsilon_0}, \ n_y^2 = \frac{\varepsilon_y}{\varepsilon_0}, \ n_z^2 = \frac{\varepsilon_z}{\varepsilon_0} \tag{6-85}$$

For uniaxial anisotropy, the index ellipsoid becomes

$$\frac{x^2}{n_0^2} + \frac{y^2}{n_0^2} + \frac{z^2}{n_e^2} = 1 \tag{6-86}$$

where $n_0^2 = \dfrac{\varepsilon_x}{\varepsilon_0} = \dfrac{\varepsilon_y}{\varepsilon_0}$, $n_e^2 = \dfrac{\varepsilon_z}{\varepsilon_0}$.

8.   OPTICAL ACTIVITY

In an anisotropic medium, the waves with orthogonal linear polarizations experience different refractive indices. They are called normal modes. For some materials, the normal modes are circularly polarized instead of linearly polarized waves. The right- and left-circularly polarized waves in the medium may experience different refractive indices. These media then act naturally as polarization rotators. That is, the direction of the electric field rotates when the wave propagates in the medium. This property is called optical activity, [6] which is usually found in materials with an inherently helical character. The materials like quartz, selenium, silver, thiogallate, tellurium, tellurium oxide, and many organic materials, e.g., sugar, have the optical activity.

8.1   Influence of Optical Activity on the Fields of EM Wave

The medium that has the optical activity has its $\vec{D}$ field not only related to the $\vec{E}$ field, but also to the magnetic field. With the helical structures, a time-varying magnetic flux density $\vec{B}$ induces a circulating current that sets up an electric dipole moment proportional to $j\omega\vec{B} = -\nabla \times \vec{E}$. The medium equation for the optical activity can be described as

$$\vec{D} = \varepsilon\vec{E} + \varepsilon_0 \xi j\omega\vec{B} = \varepsilon\vec{E} - \varepsilon_0 \xi \nabla \times \vec{E} \tag{6-87}$$

where ξ is a constant. Therefore, the optically active material is a spatially dispersive medium because the field $\vec{D}(\vec{r})$ at position $\vec{r}$ is determined not only by $\vec{E}(\vec{r})$ at a single point $\vec{r}$, but also by $\vec{E}(\vec{r})$ at points $\vec{r}$ in the vicinity of $\vec{r}$ due to the dependence on the derivatives $\nabla \times \vec{E}$.

It can be shown that the two normal modes satisfying the medium equation are circularly polarized. Consider the time harmonic plane wave $\vec{E}(\vec{r}) = \vec{E}e^{j(\omega t - \vec{k}\bullet\vec{r})}$ satisfying Eq. (6-87). With $\vec{D}(\vec{r}) = \vec{D}e^{j(\omega t - \vec{k}\bullet\vec{r})}$, Eq. (6-87) becomes

$$\vec{D} = \varepsilon\vec{E} + j\varepsilon_0 \xi\vec{k} \times \vec{E} = \varepsilon\vec{E} + j\varepsilon_0 \vec{G} \times \vec{E} \tag{6-88}$$

$\vec{G} = \xi \vec{k} = G\hat{s}$ is known as the gyration vector. $\hat{s}$ is a unit vector along the gyration vector $\vec{G}$. Because $\vec{G} \times \vec{E}$ is perpendicular to $\vec{E}$, the $\vec{D}$ field obtained from Eq. (6-88) is certainly not parallel to $\vec{E}$. The vector product $\vec{G} \times \vec{E}$ can be represented by the product of an antisymmetric tensor [G] with E, so Eq. (6-88) becomes

$$\vec{D} = (\varepsilon + j\varepsilon_0 [G])\vec{E} = \varepsilon' \vec{E} \qquad (6\text{-}89)$$

Substituting the new dielectric tensor ε' into the dispersion relation between ω and $\vec{k}$ results in the following equation

$$(n^2 - n_1^2)(n^2 - n_2^2) = G^2 \qquad (6\text{-}90)$$

where n_1 and n_2 are the refractive indices that the propagating modes experience when $G = 0$. If the normal modes are propagating along the optic axes, then $n_1 - n_2 = \overline{n}$. Eq. (6-90) has the solutions satisfying

$$n^2 = \overline{n}^2 \pm G$$

The above equation thus gives two refractive indices corresponding to the left- and right-circularly polarized modes. Because these two circularly polarized modes experience different refractive indices, their phases are not the same after propagating for some distance. In addition, their phase difference varies with the propagation distance. As a result, when they combine to form a linearly polarized wave, the polarization rotates. The rotatory power is given by

$$\rho = \frac{\pi}{\lambda}(n_l - n_r)$$

Where n_l and n_r are the solved refractive indices corresponding to the left- and right-circularly polarized modes, respectively.

8.2 Wave Equation in Optically Active Medium

For the general solution, substituting Eq. (6-89) into the Maxwell's equations leads to the wave equation for $\vec{D}$ field.

$$n^2 \hat{s} \times (\hat{s} \times \varepsilon_0 \frac{1}{\varepsilon'}) \vec{D} + \vec{D} = 0. \qquad (6\text{-}91)$$

where $\hat{s}$ is the unit vector along the direction of the gyration vector $\vec{G}$ and $\frac{1}{\varepsilon'}$ is the inverse tensor of ε'.

The eigenmodes of Eq. (6-91) are

$$\vec{D} = D_x \hat{x} + D_y \hat{y}$$

where

$$D_x = \frac{1}{2}(\frac{1}{n_1^2} - \frac{1}{n_2^2}) \pm \sqrt{\frac{1}{4}(\frac{1}{n_1^2} - \frac{1}{n_2^2})^2 + (\frac{G}{n_1^2 n_2^2})^2}$$

$$D_y = -j \frac{G}{n_1^2 n_2^2}$$

The refractive indices of the eigenmodes are given by

$$\frac{1}{n^2} = \frac{1}{2}(\frac{1}{n_1^2} + \frac{1}{n_2^2}) \pm \sqrt{\frac{1}{4}(\frac{1}{n_1^2} - \frac{1}{n_2^2})^2 + (\frac{G}{n_1^2 n_2^2})^2}$$

The unit vectors $\hat{x}$ and $\hat{y}$ are along the directions of the eigenmodes in the absence of optical activity. The two eigenmodes in the optically active medium are two elliptically polarized waves that are orthogonal to each other. In the case of an isotropic medium, $n_1 = n_2$. The corresponding polarization states are right- and left-circularly polarized waves. In the case of anisotropic materials, G is usually very small, compared to $n_2^2 - n_1^2$. Then the waves are almost linearly polarized.

For the linearly polarized light propagating in the medium, this light can be decomposed into two opposite circularly polarized modes. These two circularly polarized modes then experience different refractive indices, so their phase difference varies along the propagation distance in the optically active medium. The resultant field is the combination of the two circularly polarized light. The polarization of the resultant linearly polarized light therefore varies along the propagating distance, as shown in Fig. 6-28.

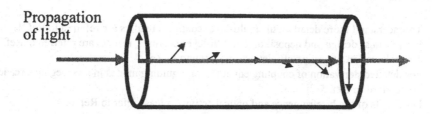

Optically active medium

Figure 6-28. Rotation of the polarization for the wave propagating in a medium with optical activity.

8.3 Faraday Rotator

Some materials act like polarization rotators when placed in a static magnetic field. The angle of rotation is proportional to the distance. This property is called Faraday effect. An optical component which uses the Faraday effect is called a Faraday rotator. The rotatory power of the Faraday rotator is proportional to the component of the B field in the direction of wave propagation.

$$\rho = VB$$

where V is called Verdet constant. The medium equation for the Faraday rotator is

$$\vec{D} = \varepsilon\vec{E} + j\varepsilon_0\gamma\vec{B} \times \vec{E}$$

In analogy to optical activity, $\vec{D} = \varepsilon\vec{E} + j\varepsilon_0\vec{G} \times \vec{E}$, here we have $\vec{G} = \gamma\vec{B}$. Note that for optical activity $\vec{G} = \xi\vec{k}$, so $\vec{G}$ depends on the direction of the wave vector. If the light reverses its propagation direction, the rotation of the polarization will trace its original rotation trajectory. In contrast, in the Faraday effect, $\vec{G}$ is independent of $\vec{k}$, so the reverse propagation of light does not reverse the rotation of the polarization. Therefore, this property can be used for making optical isolators.

NOTES

1. To understand more detail on the multi-layer coatings, readers are referred to Ref. 1.
2. For practical design and deposition of multi-layer coatings, readers are referred to Ref. 2.
3. The calculation of grating efficiency can be found in Ref. 4.
4. For detailed derivation of coupling equations and guiding modes in a waveguide, readers are referred to Ref. 5.
5. For details of the birefringence and optical activity, please refer to Ref. 6.

REFERENCES

1. Macleod, H. A., *Thin-Film Optical Filters*. American Elsevier, New York, 1969.
2. Rancourt, James D., *Optical Thin Films: User Handbook*. SPIE Press, 1996.
3. Pedrotti, F. L. and Pedrotti, L. S., *Introduction to Optics*. Prentice Hall, 1987.
4. Loewen, E. G., Neviere, M. and Maystre, D., Grating efficiency theory as it applies to blazed and holographic gratings. Applied Optics 1977; 16:2711-2721.
5. Saleh, B. E. A. and Teich, M. C., *Fundamentals of Photonics*. John Wiley and Sons, Inc., 1991.
6. Yariv, Amnon and Yeh, Pochi, *Optical Waves in Crystals*. John Wiley & Sons, 1984.

Chapter 7

PASSIVE COMPONENTS (I)

1. INTRODUCTION

With the knowledge of the optical principles used for passive components, we can now easily understand how passive components are built to perform the functions required by optical communications. In this chapter we will discuss some basic components that are usually used to form more complex passive components. In this chapter we will discuss:
- combiner/splitter;
- fixed and tunable filters;
- isolator;
- circulator;
- attenuator.

Various schemes can be used to fabricate each type of the above components to achieve a similar function. Different schemes may apply different optical principles. We will highlight the principles applied in a certain type of components and how these functions are achieved as we go through this chapter.

2. COMBINER/SPLITTER

As we briefly introduced in Chapter 6, combiners are used to combine two or more optical signal inputs from different paths into one output, while splitters do the opposite function. The different inputs and outputs have the same spectral range. They are used either as a part of the network or a part of some module. For example, we have a message encoded as an optical signal. When we want to deliver this message to several places at the same time, we use a 1 x N splitter so that optical signal is sent to different routes of optical fibers connected to those destinations. Another example is where a combiner/splitter is used for the formation of Mach-Zehnder interferometer, as shown in Fig. 7-1. In Mach-Zehnder interferometer, the input light is separated first to two paths using a splitter. The light in the two paths is then recombined using a combiner.

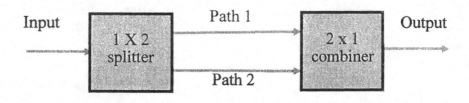

Figure 7-1. The use of combiner/splitter in a Mach-Zender interferometer.

2.1 Influence of Light's Wave Properties on Combiner/ Splitter

Although a splitter and a combiner seem to do opposite functions, the reality is not as simple as it appears. For example, a 1 x 2 splitter separates light into two paths: path one and path two. The power of light (P_{in}) is then divided to two parts. One is in path one, say P_1, and the other is in path two, say P_2. Usually we have $P_{in} = P_1 + P_2$ for many types of splitters to be discussed later in this section. However, for combiner, we can not just add up the power from two paths to the output. That is, we usually have $P_{out} \neq P_1 + P_2$, where P_{out} is the power of output light. The reason is as follows.

Because light is a wave, it has a phase. In addition, it is an EM wave, it has a property of polarization, defined as the vector direction of the electric field. When we are going to add up the power from two inputs, we actually have to add up the electric fields from the two inputs. Then the power is calculated from the square of the amplitude of the total field. Let us consider an imagined combiner that directly "combines" light from two paths. Assume that the amplitude of the field in each path is the same, while the phase could be different. Thus we have the field in path one represented as $Ee^{j\phi_1}$ and the field in path two represented as $Ee^{j\phi_2}$, where ϕ_1 and ϕ_2 are the phases when the fields from the two paths are just going to be combined. Because the power is proportional to the square of the amplitude, so the power in path one and path two is the same, $P_1 = P_2 = \beta E^2$, where β is some proportional constant. We also assume that the light from the two paths has the same polarization. The light at the output of the combiner then has the field E_{out} equal to the summation of the two fields from the two paths.

$$E_{out} = Ee^{j\phi 1} + Ee^{j\phi 2} \tag{7-1}$$

Eq.(7-1) demonstrates two interesting points. First, the field of the output light depends on the phases, $\phi 1$ and $\phi 2$. When $\phi 1$ and $\phi 2$ have a difference of π, the field of the output light is zero. Then the power of the output light is zero. When $\phi 1$ and $\phi 2$ have a difference other than π, the field of the output light is not zero and so the output power is larger than zero. Therefore, the power of the output light varies, depending on the phases, $\phi 1$ and $\phi 2$.

Second, when the two phases are the same, $\phi 1 = \phi 2$, we have $E_{out} = Ee^{j\phi 1} + Ee^{j\phi 2} = 2Ee^{j\phi 1}$. Then $P_{out} = 4\beta E^2$. However, the power in each path is only equal to βE^2. Therefore, we have $P_{out} = 4\beta E^2 > P_1 + P_2 = 2\beta E^2$. The output power is twice of total power form the inputs! How could this possibly happen? It is contradictory to the very fundamental requirement of energy conservation in physics. If there is such a combiner that makes the output power of light larger than the total power of the inputs, we will not have to worry about the lack of energy resources. The reality is that this cannot happen, so there is no such imagined combiner that directly "combines" light from two paths. In fact, a combiner is the same as a splitter or uses the same optical principles as a splitter. The detail will be discussed later for each type of combiners/splitters. In most practical cases, the output light from a combiner will have a power 3dB less than the total power of the input light. The only case where the power of the output light from the combiner could be equal to the total power of the input light is when the two inputs have orthogonal polarization of light. For example, if the first input has x-polarization of light and the second input has y-polarization of light. This will be further investigated when we discuss the polarization beam splitter.

2.2 Bulk-Type Combiner/Splitter

The bulk-type combiner/splitter is glass or other transparent dielectric material with a coating. A piece of glass or dielectric has two interfaces. In a bulk-type combiner/splitter, one of the interfaces is coated to exhibit partial reflection and partial transmission while the other interface is coated to have anti-reflection. This is shown in Fig. 7-2.

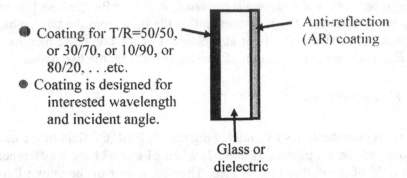

- Coating for T/R=50/50, or 30/70, or 10/90, or 80/20, . . .etc.
- Coating is designed for interested wavelength and incident angle.

Anti-reflection (AR) coating

Glass or dielectric

Figure 7-2. Bulk-type combiner/splitter.

2.2.1   Bulk-Type Splitters Using Single-Layer Coating

In devices which use a single layer of coating, this layer is sandwiched by air and the dielectric. The layer structure is shown in Fig. 7-3. Thus it behaves like the situation described in Sec. 6-2. The analysis there can then be applied to this case, so the reflection and transmission coefficients due to the single-layer coating are given by Eq. (6-11) and Eq. (6-12), respectively. The power ratios of the reflected light and the transmitted light are the absolute square of the coefficients. With the second interface of the device AR-coated, the ratio of power split can be easily calculated according to Eqs.(6-11) and (6-12).

$$R = |r|^2 \tag{7-2}$$

$$T = |t|^2 \tag{7-3}$$

where r is given by Eq.(6-11) and t is given by Eq. (6-12). Therefore, the input light is divided into two outputs. One output has R x 100 % of the input power and the other has T x 100 % of the input power. Here we assume that the output side is also in air. If the output is not in the same medium as the input, the power ratio of the transmitted power is (1-R).

In order to easily extract the output light from the splitter, the input light is usually not normal to the splitter. A common incidence angle for input light is 45°. Then the TE wave and the TM wave behave differently because their reflection and transmission coefficients are not the same, as discussed in Sec.6-1. Therefore, the bulk-type splitter with a single-layer coating is polarization dependent. In addition, coating using a single layer only works for a small spectral range. Therefore, single-layer coatings are seldom used for bulk-type splitters.

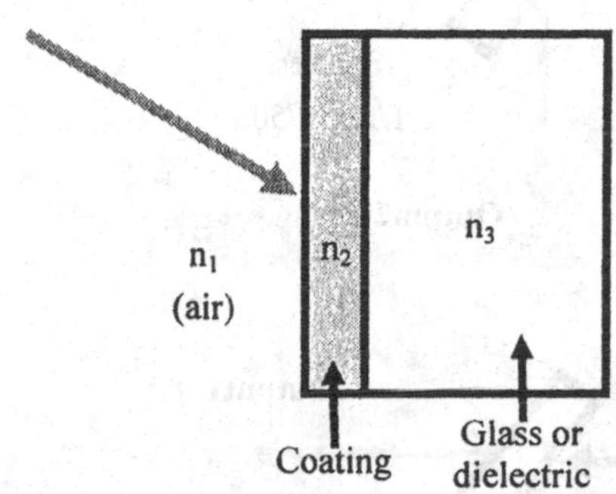

Incident light

n_1
(air) n_2 n_3

Coating   Glass or dielectric

Figure 7-3. Layer structure of a single-layer coating for bulk-type combiner/splitter.

2.2.2   Bulk-Type Splitters Using Multi-Layer Coatings

Reflection and transmission from multi-layer coatings was discussed in Sec. 6-3. Because multi-layer coatings provide a variety of selections on the refractive index and the layer thickness, many kinds of combinations for the percentage of reflection and transmission can be achieved. It is possible to have a certain percentage of reflection and transmission over a broad bandwidth. Although the reflection and transmission coefficients for TE wave and the TM wave are not the same, according to Eq. (6-52) and Eq. (6-53), it is still possible to make the splitter polarization insensitive over a limited spectral range. So even though single layer coatings are much simpler to fabricate, coating with multi-layer dielectrics is more commonly used.

Fig. 7-4 shows two examples of the bulk-type splitter. In Fig.7-4 (a) the reflection of power is equal to the transmission of power. Thus the input is divided to two outputs. Each output has half of the input power. In Fig. 7-4(b), the transmitted light has 80 % of the input power and the reflected light has 20 % of the input power.

2.2.3   Phase Difference between the Two Outputs from the Bulk-Type Splitter

When we are only interested in the power of the two outputs from the splitter, the phase is not important. However, because a splitter is usually not used alone in the optical communication network, other components quite possibly pick up the optical signals from the outputs of a splitter. The phase of each splitter output may possibly influence the behavior of light in the latter components. Therefore, it is worthwhile to look at the phase of the

(a)

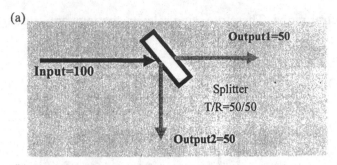

(b)

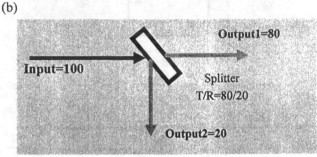

Figure 7-4. Two examples of bulk-type splitter: (a) equal ratio of power splitting; (b) input power is divided to the ratio of 8:2.

outputs from the splitter. Because the outputs of the bulk-type splitter are the reflected light and the transmitted light, we will investigate the phase in the transmission and reflection coefficients.

Usually only one interface of the splitter is coated for the purpose of beam splitting, while the other interface is AR-coated. There are two possible situations to use the splitter. First, the first interface that light encounters is used for beam splitting, while the second interface is AR-coated. The situation is illustrated in Fig. 7-5 (a). In this case, light enters the splitter from the air. The reflected beam goes back to air directly, while the transmitted light is in the glass or dielectric. The second case is shown in Fig. 7-5 (b). Light first passes through the AR-coated interface and then gets split at the second interface. Then, the reflected beam goes back to the glass or dielectric, while the transmitted light is in the air right after the splitting action . Phase difference between the two outputs is not the same for the above two situations. They will be analyzed separately in the following.

For the situation shown in Fig. 7-5(a), we can directly apply the results of multi-layer coatings derived in Sec. 6-3. Let us rewrite the reflection and transmission coefficients here. $r = \dfrac{n_0 M_{11} + n_0 n_s M_{12} - M_{21} - n_s M_{22}}{n_0 M_{11} + n_0 n_s M_{12} + M_{21} + n_s M_{22}}$ and

$t = \dfrac{2n_0}{n_0 M_{11} + n_0 n_s M_{12} + M_{21} + n_s M_{22}}$. As shown in Fig. 7-5(a), n_0 is the refractive index of air and n_s is the refractive index of the glass or dielectric. The two coefficients are related to the elements of the characteristic matrix of the multi-layer coatings $M = M_1 M_2 M_3 \cdots M_m = \begin{bmatrix} M_{11} & M_{12} \\ M_{21} & M_{22} \end{bmatrix}$, where $M_1, M_2, \ldots, M_m$ are the characteristic matrices of those layers. For inclined incidence, the refractive index of each layer is replaced either by Eq.(6-52) for TM wave or Eq.(6-53) for TE wave. Therefore, the ratio of the reflection coefficient to the transmission coefficient is given by

$$\frac{r}{t} = \frac{n_0 M_{11} + n_0 n_s M_{12} - M_{21} - n_s M_{22}}{2n_0} \tag{7-4}$$

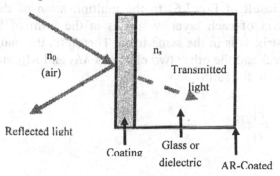

(a) Incident light

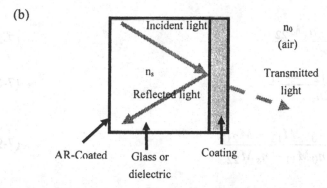

(b)

Figure 7-5. Use of a bulk-type splitter: (a) light enters the coating layer (for splitting) from the air; (b) light enters the coating layer (for splitting) from the glass or dielectric.

To determine the phase of the ratio r/t, we have to investigate the properties of those matrix elements M_{11}, M_{12}, M_{21}, and M_{22}. Because the matrix is equal to the multiplication of the individual characteristic matrix of each layer, we have to know the properties of the matrix elements for each individual matrix. From Eq.(6-43), we know that each individual matrix is in the following form

$$M_i = \begin{bmatrix} \cos\phi_i & j\sin\phi_i/n_i \\ jn_i\sin\phi_i & \cos\phi_i \end{bmatrix} = \begin{bmatrix} A & jB \\ jC & D \end{bmatrix} \quad i = 1, 2, 3, \ldots, m \quad (7\text{-}5)$$

where A, B, C, D are real numbers. Eq. (7-5) means that the matrix elements M_{i11} and M_{i22} are real, while the other two elements M_{i12} and M_{i21} are imaginary. It can be easily shown that the matrix product of two matrices both in the form of Eq. (7-5) is still in the same form.

$$\begin{bmatrix} A_1 & jB_1 \\ jC_1 & D_1 \end{bmatrix}\begin{bmatrix} A_2 & jB_2 \\ jC_2 & D_2 \end{bmatrix} = \begin{bmatrix} A_3 & jB_3 \\ jC_3 & D_3 \end{bmatrix} \tag{7-6}$$

Applying the result of Eq.(7-6) to the multiplication of the individual characteristic matrix of each layer, which is in the form of Eq.(7-5), we obtain that the matrix M is in the same form. Therefore, the matrix elements M_{11} and M_{22} are real and the other two elements M_{12} and M_{21} are imaginary. The ratio r/t is then in the form

$$\frac{r}{t} = X + jY = \left|\frac{r}{t}\right|e^{j\alpha} \tag{7-7}$$

where

$$X = \frac{n_0 M_{11} - n_s M_{22}}{2n_0} \tag{7-8a}$$

$$jY = \frac{n_0 n_s M_{12} - M_{21}}{2n_0} \tag{7-8b}$$

$$\tan\alpha = \frac{Y}{X} = \frac{n_0 n_s M_{12} - M_{21}}{n_0 M_{11} - n_s M_{22}} \tag{7-9}$$

The phase difference between the reflected light and the transmitted light is thus obtained using by Eqs (7-8a), (7-8b), and (7-9).

For the situation shown in Fig. 7-5(b), we can also directly apply the results of multi-layer coatings derived in Sec. 6-3 with the reverse order of the matrix labeling. However, in order to make easy comparison with the results shown in Eqs (7-8a), (7-8b), and (7-9), we will keep the same order of matrix labeling. That is, the layer with the same characteristic matrix M_l as the top-most (next to air) and the last layer with characteristic matrix M_m nearest the glass or the dielectric. The layer structure of coating is shown in

Fig. 7-6. The order of the layers is the same for the situations shown in Fig.7-5 (a) and 7-5(b). The conventions of those symbols are the same as those defined in Sec.6-3 so that the mathematical derivations there can be applied. Using Eqs. (6-45), (6-46) and (6-47), we obtain the relation for $E^-(Z_1^-)$, $E^+(Z_1^-)$, $E^-(Z_{m+1}^+)$ and $E^+(Z_{m+1}^+)$.

$$\begin{bmatrix} E^+(Z_1^-) \\ E^-(Z_1^-) \end{bmatrix} = \begin{bmatrix} \dfrac{1}{2} & \dfrac{1}{2n_0} \\ \dfrac{1}{2} & -\dfrac{1}{2n_0} \end{bmatrix} \begin{bmatrix} M_{11} & M_{12} \\ M_{21} & M_{22} \end{bmatrix} \begin{bmatrix} 1 & 1 \\ n_s & -n_s \end{bmatrix} \begin{bmatrix} E^+(Z_{m+1}^+) \\ E^-(Z_{m+1}^+) \end{bmatrix}$$

$$(7\text{-}10)$$

where the matrix $M = M_1 M_2 M_3 \cdots M_m - \begin{bmatrix} M_{11} & M_{12} \\ M_{21} & M_{22} \end{bmatrix}$ is the same as the one given in Sec. 6-3.

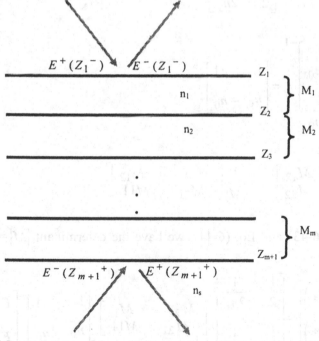

$E^+(Z_1^-)$ $E^-(Z_1^-)$ Z_1 $\Big\}$ M_1

n_1 Z_2

n_2 Z_3 $\Big\}$ M_2

$\Big\}$ M_m

Z_{m+1}

$E^-(Z_{m+1}^+)$ $E^+(Z_{m+1}^+)$

n_s

Figure 7-6. Layer structure of the multi-layer coatings. The conventions of the symbols are the same as those defined in Sec. 6-3.

For the reflection coefficient derived in Sec. 6-3 and the situation shown in Fig. 7-5(a), $E^-(Z_{m+1}^+)=0$ because there is no reflected wave after the last boundary $z = Z_{m+1}$. Now we want to calculate the reflection coefficient and the transmission coefficient for the light incident from the glass or the dielectric to the multi-layer coating. The relation in Eq.(7-10) is rewritten to the following formula.

$$\begin{bmatrix} E^+\left(Z_{m+1}^+\right) \\ E^-\left(Z_{m+1}^+\right) \end{bmatrix} = \begin{bmatrix} 1 & 1 \\ n_s & -n_s \end{bmatrix}^{-1} \begin{bmatrix} M_{11} & M_{12} \\ M_{21} & M_{22} \end{bmatrix}^{-1} \begin{bmatrix} \dfrac{1}{2} & \dfrac{1}{2n_0} \\ \dfrac{1}{2} & -\dfrac{1}{2n_0} \end{bmatrix}^{-1} \begin{bmatrix} E^+\left(Z_1^-\right) \\ E^-\left(Z_1^-\right) \end{bmatrix}$$

$$(7\text{-}11)$$

We have

$$\begin{bmatrix} 1 & 1 \\ n_s & -n_s \end{bmatrix}^{-1} = \begin{bmatrix} \dfrac{1}{2} & \dfrac{1}{2n_s} \\ \dfrac{1}{2} & -\dfrac{1}{2n_s} \end{bmatrix}$$

$$\begin{bmatrix} \dfrac{1}{2} & \dfrac{1}{2n_0} \\ \dfrac{1}{2} & -\dfrac{1}{2n_0} \end{bmatrix}^{-1} = \begin{bmatrix} 1 & 1 \\ n_0 & -n_0 \end{bmatrix}$$

$$\begin{bmatrix} M_{11} & M_{12} \\ M_{21} & M_{22} \end{bmatrix}^{-1} = \dfrac{1}{|M|}\begin{bmatrix} M_{22} & -M_{12} \\ -M_{21} & M11 \end{bmatrix}$$

From Eq.(6-43) and Eq. (6-44), we have the determinant $|M|=1$. Eq.(7-11) becomes

$$\begin{bmatrix} E^+\left(Z_{m+1}^+\right) \\ E^-\left(Z_{m+1}^+\right) \end{bmatrix} = \begin{bmatrix} \dfrac{1}{2} & \dfrac{1}{2n_s} \\ \dfrac{1}{2} & -\dfrac{1}{2n_s} \end{bmatrix} \begin{bmatrix} M_{22} & -M_{12} \\ -M_{21} & M11 \end{bmatrix} \begin{bmatrix} 1 & 1 \\ n_0 & -n_0 \end{bmatrix} \begin{bmatrix} E^+\left(Z_1^-\right) \\ E^-\left(Z_1^-\right) \end{bmatrix}$$

$$(7\text{-}12)$$

For the situation shown in Fig. 7-5(b), the incident wave is $E^-(Z_{m+1}^+)$, while the reflected wave and the transmitted wave are $E^+(Z_{m+1}^+)$ and

$E^-(Z_1^-)$, respectively. $E^+(Z_1^-)$ represents the reflected wave in air for this case, so $E^+(Z_1^-)=0$ because there is no reflected wave in air. The reflection coefficient (r) and the transmission coefficient (t) are then as follows.

$$r = \frac{E^+(Z_{m+1}^+)}{E^-(Z_{m+1}^+)} = \frac{-n_0 M_{11} + n_0 n_s M_{12} - M_{21} + n_s M_{22}}{n_0 M_{11} + n_0 n_s M_{12} + M_{21} + n_s M_{22}} \tag{7-13}$$

$$t = \frac{E^-(Z_1^-)}{E^-(Z_{m+1}^+)} = \frac{2n_s}{n_0 M_{11} + n_0 n_s M_{12} + M_{21} + n_s M_{22}} \tag{7-14}$$

The ratio of the reflection coefficient to the transmission coefficient is now given by

$$\frac{r}{t} = \frac{-n_0 M_{11} + n_0 n_s M_{12} - M_{21} + n_s M_{22}}{2n_s} \tag{7-15}$$

This ratio can be further given in the form

$$\frac{r}{t} = -X' + jY' = \left|\frac{r}{t}\right| e^{j\alpha'} \tag{7-16}$$

where

$$X' = \frac{n_0 M_{11} - n_s M_{22}}{2n_s} \tag{7-17a}$$

$$jY' = \frac{n_0 n_s M_{12} - M_{21}}{2n_s} \tag{7-17b}$$

$$\tan \alpha' = -\frac{Y'}{X'} = -\frac{n_0 n_s M_{12} - M_{21}}{n_0 M_{11} - n_s M_{22}} \tag{7-18}$$

The phase difference between the reflected light and the transmitted light is now obtained using Eqs (7-17a), (7-17b), and (7-18).

Comparing Eq. (7-7) and Eq. (7-16), we have

$$\alpha' + \alpha = \pi \tag{7-19}$$

This indicates that the phase difference is not the same for light incident from the air or the glass. The situation is illustrated in Fig. 7-7. Beam one is incident from the air, while beam two is incident from the glass. r_1 and t_1 are, respectively, the reflected field and the transmitted field of beam one. Also, r_2 and t_2 are, respectively, the reflected field and the transmitted field of beam two.

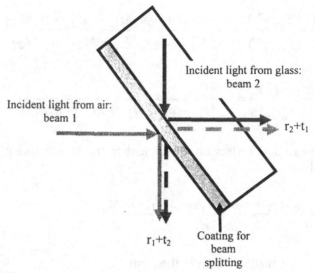

Figure 7-7. Reflection and transmission of beams at the splitter. Beam one is incident from the air, indicated by the orange arrow lines. Beam two is incident from the glass, indicated by red arrow lines. Dashed arrows indicate transmitted beams. The summation of r_1+t_2 and r_2+t_1 will have different interference effects due to the various phase differences discussed in the text.

We have $\dfrac{r_1}{t_1} = \left|\dfrac{r_1}{t_1}\right| e^{j\alpha}$ and $\dfrac{r_2}{t_2} = \left|\dfrac{r_2}{t_2}\right| e^{j\alpha'}$ with α and α' given in Eq. (7-7) and Eq. (7-16), respectively. α and α' satisfy Eq. (7-19). The sum of r_1+t_2 and r_2+t_1 at the two outputs will have different behaviors even when the power is equally divided between the two outputs. Therefore, the phase difference between the reflected light and the transmitted light does cause significant effects for some cases. The detail will be further discussed in Sec. 7.2.2.4.

2.2.4   Considerations for a Combiner

As we mentioned before, there is no combiner that directly "combines" light from two inputs. In fact, a combiner is also a glass or other transparent dielectric with one interface coated to have partial reflection and transmission and the other interface coated for anti-reflection, as shown in

Fig. 7-2. Therefore, its basic function is the same as a splitter. The difference is in its use. When it is used as a splitter, only one beam of light is delivered to the device. When it is used as a combiner, two or more beams of light are incident on the device from different directions.

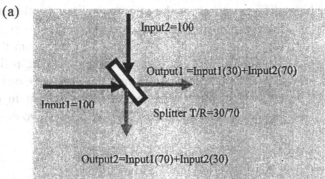

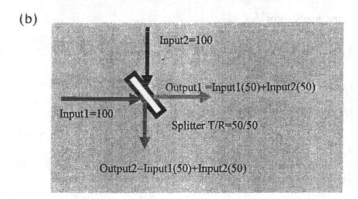

Figure 7-8. A splitter used as a combiner: (a) splitter with T/R=30/70; (b) splitter with T/R=50/50.

Two examples of using such a device as a combiner are shown in Fig. 7-8. There are always two outputs. In Fig. 7-8 (a), the splitter has T/R = 30/70, so 30% of power in input one and 70% of power in input two go to output one. In addition, 70% of power in input one and 30% of power in input two go to output two. If we use output one only, we combine 30% of power in input one and 70% of power in input two with this device. On the other hand, if we are interested in output two only, this device works as a combiner that combines 70% of power in input one and 30% of power in input two. In Fig. 7-8 (b), the splitter has T/R = 50/50, so 50% of power in input one and 50% of power in input two go to output one. Also, 50% of power in input one and 50% of power in input two go to output two. It looks as if both output one and output two have the same result. However, if we consider the phase difference discussed between the reflected light and the transmitted light before, we will see that output one and output two behave in an opposite way due to the interference of the reflected light and the transmitted light. The detailed analysis is in the following.

Consider the case shown in Fig. 7-8(b). We assume that input one and input two have the same power. The light in input one and the light in input two have a phase difference before they hit the splitter/combiner. Then we can express the field in input one and the field in input two as a and a $\exp(j\phi_i)$, respectively. Thus the reflected field and the transmitted field of light from input one are ar_1 and at_1, respectively. Also, the reflected field and the transmitted field of light from input two are $ar_2 \exp(j\phi_i)$ and at_2 $\exp(j\phi_i)$, respectively. Therefore, we have the field of light in output one equal to $[at_1 + ar_2 \exp(j\phi_i)]$ and the field of light in output two equal to $[ar_1 + at_2 \exp(j\phi_i)]$.

$$E_{o1} = at_1 + ar_2 \exp(j\phi_i) \tag{7-20a}$$

$$E_{o2} = ar_1 + at_2 \exp(j\phi_i) \tag{7-20b}$$

According to Eq.(7-7) and Eq. (7-16), we have

$$r_1 = \left|\frac{r_1}{t_1}\right| t_1 e^{j\alpha_1} = \kappa_1 t_1 e^{j\alpha_1} \tag{7-21a}$$

$$r_2 = \left|\frac{r_2}{t_2}\right| t_2 e^{j\alpha_2} = \kappa_2 t_2 e^{j\alpha_2} \tag{7-21b}$$

where $\alpha_1 + \alpha_2 = \pi$ according to Eq.(7-19). In addition, from the comparison of Eq.(6-49) and Eq.(7-14), we have $\dfrac{t_1}{t_2} = \dfrac{n_0}{n_s}$. The reason that the ratio of the two transmission coefficients are not equal to one is because the power ratio has to take into account the dielectric constant in each dielectric, as shown in Eq. (6-51). Substituting r_1 and r_2 into Eqs.(7-20a) and (7-20b), we obtain

$$E_{o1} = at_1 + a\kappa_2 t_2 e^{j(\alpha_2 + \phi_i)} = at_1[1 + \frac{n_s}{n_0}\kappa_2 e^{j(\alpha_2 + \phi_i)}] \tag{7-22a}$$

$$E_{o2} = a\kappa_1 t_1 e^{j\alpha_1} + at_2 e^{j\phi_i} = at_1[\kappa_1 e^{j\alpha_1} + \frac{n_s}{n_0}e^{j\phi_i}] \tag{7-22b}$$

The power is proportional to the absolute square of the above quantities. Thus

$$P_{o1} \propto |at_1|^2 [1 + (\frac{n_s}{n_0} \kappa_2)^2 + 2 \frac{n_s}{n_0} \kappa_2 \cos(\alpha_2 + \phi_i)] \qquad (7\text{-}23a)$$

$$P_{o2} \propto |at_1|^2 [\kappa_1^2 + (\frac{n_s}{n_0})^2 + 2 \frac{n_s}{n_0} \kappa_1 \cos(-\alpha_1 + \phi_i)]$$

$$= |at_1|^2 [\kappa_1^2 + (\frac{n_s}{n_0})^2 + 2 \frac{n_s}{n_0} \kappa_1 \cos(\alpha_2 + \phi_i + \pi)] \qquad (7\text{-}23b)$$

Taking into account the effect of the dielectric constant on the calculation of power at the AR-coated interface, we have the powers at the two outputs in the air given by

$$P_{o1} \propto [1 + \cos(\alpha_2 + \phi_i)] \qquad (7\text{-}24a)$$

$$P_{o2} \propto [1 + \cos(\alpha_2 + \phi_i + \pi)] \qquad (7\text{-}24b)$$

Fig. 7-9 shows the variation of the powers of output one and output two with the phase difference between the light in the two inputs, ϕ_i. We see that when the power at output one is maximum, then the output power at output two is minimum. In this way, we have $P_1 + P_2 = P_{in}$, so the energy is conserved.

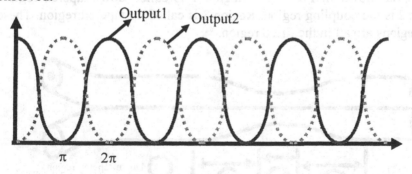

Output1   Output2

π 2π

Figure 7-9. Variation of the powers of the output1 and the output2 with the phase difference between the light in the two inputs, ϕ_i.

2.2.5   Package of Bulk-Type Combiner/Splitter

The bulk-type combiner/splitter is used for light traveling in free space. However, for optical fiber communication, light is propagating in the fiber. It is not convenient to use bulk-type combiners/splitters if they are not properly packaged. Fig. 7-10 shows an example of package for this type of splitters. Because the aperture of a fiber is small, as light leaves the fiber, it will diverge. Therefore, input light is collimated using a micro lens. After it passes the splitter, it is focused to the fiber center using another micro lens.

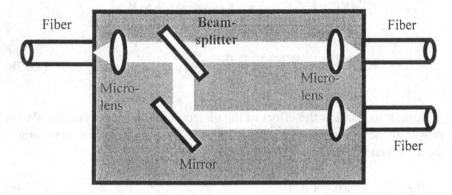

Figure 7-10. An example of package for bulk-type splitter.

2.3   Fused-Fiber Combiner/Splitter

Fused-fiber type combiners/splitters are based on optical fibers, so it is very convenient to use them in the fiber communication system. Fig. 7-11 shows a schematic of this type of combiner/splitter. It can be divided to three, regions, as shown in Fig. 7-11. Region 1 is called down-tapered region. Region 2 is the coupling region. Region 3 is called up-tapered region. These three regions are all in the fused region.

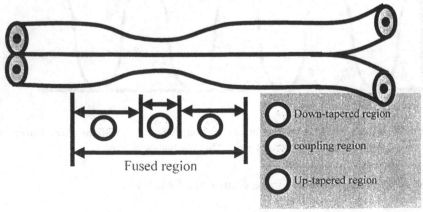

Figure 7-11. Schematic of fused-fiber combiner/splitter.

Fused-fiber combiners/splitters utilize the wave coupling in the waveguide, which is the fiber here. The coupling of wave from one waveguide to another is explained in Chapter 6. For the coupling to happen, the two waveguides have to be very close, so there is modal overlap and the coupling coefficients given in Eqs. (6-67a) and (6-67b) are not zero. However, the core of the single-mode fiber has a diameter of only 9 μm. This core is surrounded by a thick cladding and the wave is mainly confined within the core. It is not possible to have coupling of waves between the fibers by just putting together two fibers side by side. In order to make the coupling happen, the core in one fiber has to be very close to the core of another fiber or the wave in the fiber has to extend far outside the core.

A common way to achieve this is that two single-mode fibers are melted and pulled together. Then a fused region that consists of the tapered-down region, the coupling region, and the up-tapered region shown in Fig 7-11 is formed. In the tapered-down region, the cross sections of the two fibers are gradually reduced to the size in the coupling region. In the coupling region, the dimension of the fiber is very small, so the V number (proportional to a/λ, where a is the diameter of the core) is very small. This causes the wave guided in the fiber to extend far outside the core and the transverse profile of the wave to possibly extend near the core of another fiber. As a result, coupling occurs according to the coupling theory described in Chapter 6. By properly controlling the dimension of fiber in the coupling region, a desired ratio of power coupling can be obtained. In practical fabrication, light is launched from one of the inputs. The optical powers from the two outputs are monitored during the heating and pulling process. After the coupling region, the waves move through the up-tapered region and then separate. Thereafter they propagate in two separate fibers. Due to the taper region, when the wave is shaped from the fiber to the coupling region and then reshaped to the fiber, there is excess loss. This loss may be small however. Typical excess loss is from 0.05 dB to 0.2 dB. The coupling ratio from 50:50 to 1:99 can be fabricated with good control.

Shown in Fig. 7-11 is a 2 x 2 fused-fiber combiner/splitter. There are two input ends and two output ends, so it is possible to use it as a splitter or a combiner. If only one input is used, it works as a splitter. Fig. 7-12 shows two examples of wave coupling from one fiber to another in the fused-fiber splitter. If both inputs are used, it is used as a combiner.

When it is used as a combiner, two waves are launched into the two fibers separately at the two input ends. The wave in the top fiber will couple to the bottom one in the coupling region. At the same time, the wave in the bottom fiber will also couple to the top waveguide. If the two fibers are identical in the coupling region, both waves will couple the same ratio of power from one fiber to another. Even the two fibers are not identical in the coupling region, there is also some ratio of power coupling from one fiber to another for both input waves. Therefore, it is not possible to completely combine the power of the two input waves to only one output, similar to the effect we discussed for bulk-type combiners. Similar consideration is also necessary for the phase effect on the power ratio between the two outputs.

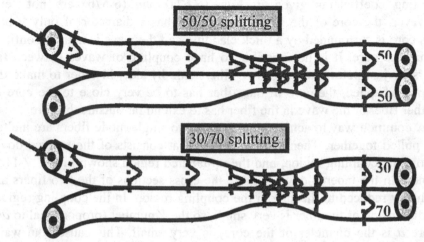

Figure 7-12. Wave coupling in the fused-fiber splitter. Top figure: 50/50 coupling; bottom figure: 30/70 coupling.

2.4   Waveguide Combiner/Splitter

Waveguide combiners/splitters also utilize wave coupling in the waveguide. The details of wave coupling are explained in Chapter 6. Fig. 7-13 shows a schematic of a waveguide combiner/splitter. In this figure, a buried waveguide is used. Ridge waveguides are also commonly used for waveguide combiners/splitters. The selection of waveguide types depends on the fabrication techniques used for the substrate of the waveguide. For example, if LiNbO$_3$ is used, a buried waveguide is often used because the etching of LiNbO$_3$ is difficult. On the other hand, SOI (Si on insulator) has gained in popularity for its easy fabrication using the mature Si technology and its transparency in the 1.3 μm and 1.55 μm wavelength regions. When SOI is used, a ridge waveguide can be easily fabricated, so a ridge waveguide is adopted.

Again, for coupling to happen, the two waveguides have to be in close proximity to cause the modes in the two waveguides to overlap, so the coupling coefficients given in Eqs. (6-67a) and (6-67b) are not zero. It is easy to fabricate two waveguides in close proximity using current lithography techniques. However, we have to deliver light into and out of the waveguides. When the two waveguides are very close, it is difficult to couple light to each individual waveguide. Therefore, we fabricate the waveguide with wide separation outside the coupling region. In Fig. 7-13, the wide spacing of the two waveguides is changed to close spacing in the coupling region using sharp folded angles. The spacing of the two waveguides can also be gradually changed using a bending structure. The change of the waveguide direction will induce excess loss that is about 0.3-0.5 dB.

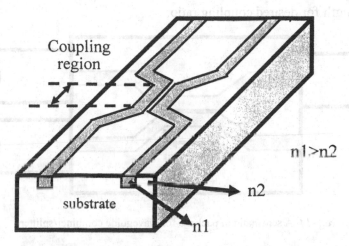

Figure 7-13. Schematic of waveguide combiner/splitter.

It is also not convenient to directly use the waveguide combiners/splitters in fiber-optic communications. Since both the core of the fiber and the cross section of the waveguide are very small, it is difficult for general users to align them in a straight line for light delivery from the fiber to the waveguide or vice versa. Prior fabrication of a package with fibers already connected to the waveguide combiner/splitter is therefore necessary. Fig. 7-14 schematically shows a top view of the package. In this figure, the light in the fiber is directly coupled to the waveguide without mode shaping. If the cross section of the waveguide in the waveguide combiner/splitter is highly asymmetric, some mode shaping may be necessary, like the coupling of light between the fiber and the semiconductor optical amplifier. The coupling loss of the single-mode fiber is about 0.3-1 dB. Although it requires additional effort for packaging when compared to the fused-fiber combiner/splitter, its fabrication cost is much cheaper. Because it is fabricated using semiconductor processing techniques, mass-production is possible.

Shown in Fig. 7-14 is a 2 x 2 waveguide combiner/splitter. There are also two input ends and two output ends, so it is possible to use it as a splitter or a combiner. If only one input is used, it works as a splitter. If both inputs are used, it is used a combiner. The consideration for its use as a combiner is similar to previous discussions on other types of combiners/splitters.

As we discussed in Chapter 6, to control the coupling ratio of power, we can design the coupling length or the phase mismatch $\Delta\beta$. Because the waveguide combiners/splitters are fabricated using semiconductor processing techniques, it is particularly convenient to control those parameters. Fig. 7-15 shows the coupling ratio vs. the coupling length. The repeatability of semiconductor processing makes it possible to control the coupling length for desired coupling ratio.

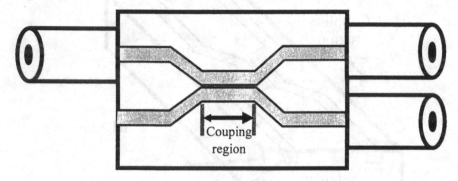

Figure 7-14. A schematic of package for waveguide combiner/splitter.

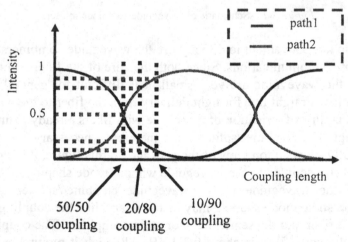

Figure 7-15. Coupling ratio vs. coupling length.

2.5   Wavelength Dependence of Fused-Fiber Combiner/Splitter and Waveguide Combiner/Splitter

According to the analysis in Chapter 6, the coupling length depends on the coupling coefficients. From Eqs. (6-67a) and (6-67b), we know that the coupling coefficients are functions of the wavelength through several factors.

First, the wave vectors k_o, β_1 and β_2 have strong dependence on the wavelength. Second, the field distribution of the guiding modes $u_1(y)$ and $u_2(y)$ have slight variation with the wavelength. Third, material dispersion causes the refractive index to slightly vary with the wavelength. The latter two factors are usually not important. However, the first factor could make the coupling ratio vary significantly with the wavelength as the coupling length is fixed.

In many applications, we would like to have wavelength-independent combiners/splitters. Several methods have been used to reduce the wavelength dependence of fused-fiber combiners/splitters. They include
- the use of fibers with different diameters;
- the use of fibers with different refractive indices;
- The use of different tapered shapes.

A coupling ratio of 50% ± 9% can be achieved for the range of wavelengths from 1.23 μm to 1.57 μm.[1]

For waveguide combiners/splitters, similar concepts can be applied, e.g., using two waveguides of different widths, doping the buried waveguides differently, gradual change of the width of the input waveguide, and so on. A uniformity of ± 0.6 dB or 20% ± 1.9% of the coupling ratio over the range of wavelengths from 250 nm to 1650 nm can be obtained using various schemes.[2, 3]

2.6 Y-Couplers

The above fused-fiber combiners/splitters and the waveguide combiners/ splitters using guiding mode coupling from one waveguide to another are also called directional couplers. There is another type of coupler simpler than those above called an Y-coupler. Its schematic is shown in Fig. 7-16. . The waveguide can be constructed using optical fibers or waveguides fabricated on some substrate. The splitting of light from the input to two

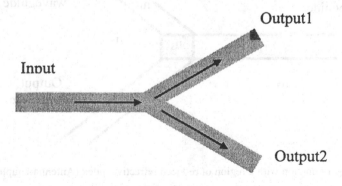

Figure 7-16. A schematic of Y-coupler.

outputs is quite straightforward. There is large radiation loss at the junction region however because the guiding mode experiences a significant change in the guiding region. To reduce this loss, some modification on the Y-junction is used. Fig. 7-17 shows that the input waveguide is expanded with a taper region before it is split. The guiding mode then gradually expands to a larger region, making the mode better at splitting into the two latter waveguides.

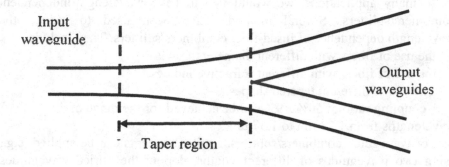

Figure 7-17. A Y-junction with taper region.

In Fig. 7-18, a region with a smaller refractive index is introduced to the end of the input waveguide. $n_3 < n_2$ and $n_3 < n_1$, where n_1 is the refractive index in the waveguide, n_2 is the refractive index outside the waveguide and is the refractive index of the additionally introduced region. This structure is called antenna-coupled Y-coupler. When the guiding mode in the input waveguide propagates into this region, the mode expands due to a lack of guiding effect of the reduced refractive index. This makes the wave front of

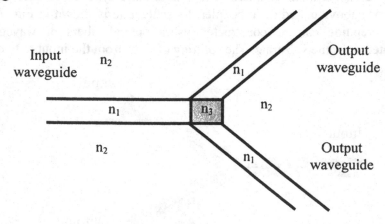

Figure 7-18. A Y-junction with a region of reduced refractive index (Antenna-coupled Y-coupler).

the guiding mode curved and so it can be perpendicular to the latter waveguides for outputs. Thus this mode can be guided in the output waveguides much better than the situation in the simple Y-junction, giving rise to reduced loss. A usual Y-junction without special treatment typically experiences loss of no less than 3 dB. The modification using a region with a reduced refractive index could reduce the loss to only 1 dB.[4]

2.7   1 X N Couplers and N x N couplers

Using the 1 x 2 or 2 x 2 combiners/splitters described above, we can construct 1 x N couplers or N x M couplers. The 2 x 2 combiner/splitter with the power ratio of 50/50 is usually used. Because the power is divided in half for each output, it is also called 3 dB coupler. Fig. 7-19 shows an example of using four 1 x 2 waveguide combiners/splitters to form a 1 x 4 coupler.

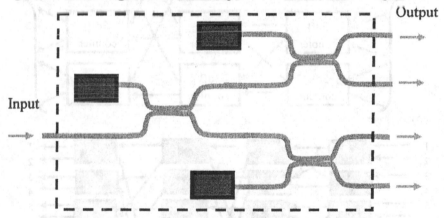

Figure 7-19. A 1 x 4 coupler formed by three waveguide combiners/splitters.

To form an N x N coupler, we require N_0 number of 3 dB combiners/splitters.

$$N_0 = \frac{N}{2} \log_2(N) \tag{7-25}$$

For example, the number of 3 dB couplers required to form a 4 x 4 coupler is 4, while the number required to form 8 x 8 is 12. For a 16 x 16 coupler, we need thirty-two 3 dB couplers. The configuration of the 8 x 8 and 16 x 16 couplers is shown in Fig. 7-20. The bulk-type, fused-fiber, and the waveguide 2 x 2 combiners/splitters can all be used to form N x N couplers. Usually fused-fiber combiners/splitters are used for low-cost and low-port-count applications. Because the assembly process takes time, they are not used for high-port count applications. In contrast, waveguide combiners/splitters can be monolithically integrated on a substrate using semiconductor processing techniques, so they are more suitable for applications over 16 ports than fused-fiber types. For high-performance communication systems, bulk-type combiners/splitters packaged with micro-optics are used.

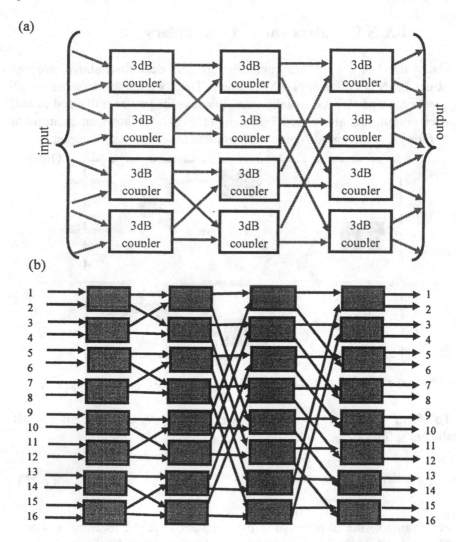

Figure 7-20. (a) A 8 x 8 coupler using twelve 3dB couplers. (b) A 16 x 16 coupler using thirty-two 3dB couplers. The red rectangular box represents a 3 dB coupler.

For a 1 x N coupler or an N x N coupler, the input signals travel through $\log_2(N)$ 3dB couplers. The input power of optical signals is divided to each output. Ideally each output will share a fraction 1/N of the input power.

However, there is inevitably excess loss for each 3 dB coupler. Assume that each 3 dB coupler has the excess loss of α ($0< \alpha <1$). That is, the signal before the 3 dB coupler has a power reduced to α times of the power before it enters the 3 dB coupler. As the signals travel from the input to the output, their power will be reduced to $\alpha^{\log_2(N)}$. Therefore, the power at each output is $\alpha^{\log 2(N)}/N$ times of the input power. The excess loss is loss introduced by the 3 dB couplers, usually expressed in dB.

$$Excess\ loss = 10\log_{10}[\alpha^{\log(N)}] = 10[1 - 3.3\log_{10}(\alpha)]\log_{10}(N) \quad (7\text{-}25)$$

With proper fabrication, this excess loss can be held to small values. [5]

The multi-port couplers can also be fabricated without using the 2 x 2 combiners/splitters. For example, 8 fibers may be bundled, twisted, and then heated and pulled to form an 8 x 8 star coupler. The fabrication process is like the formation of a 2 x 2 fused-fiber combiners/splitters. [6] However, because the coupling behaviors among those fibers can not be easily controlled, it is difficult to fabricate star couplers with high-port count using this method. Another way to form star couplers is to fabricate an array of input waveguides and output waveguides that are monolithically integrated with a planar region. The planar region is placed between the input waveguides and the output waveguide. It behaves like free space. Thus when the wave from the input waveguide enters this region, it spreads out and then is received by each output waveguide. Proper design of the planar region is thus necessary in order to make the power split to those output waveguides with good port-to-port uniformity. Many ports of star couplers with low excess loss and good uniformity of power splitting among the output ports have been reported. [7]

In the N x M star coupler, the signals from one input are transmitted to all outputs at the same time. If all of inputs have signals, they are mixed and transmitted to the outputs. Therefore, it works like a radio communication system in free space. It then has the disadvantages of a radio communication system like severe crosstalk due to the mixed input signals and the reduced power of signals at the receiver.

2.8 Polarization Beam Splitters

The above couplers or splitters usually divide the input power of optical signals into outputs according to the desired power ratio regardless of the polarization of input light. A polarization beam splitter however divides the input light to the outputs according to its polarization. As shown in Fig. 7-21, the input light has two orthogonal polarizations. The interface for beam splitting is sandwiched in the cube. The horizontally-polarized p-wave (TM wave) transmits through the interface for beam splitting, while the vertically-polarized s-wave (TE wave) is reflected by the same interface.

The polarization beam splitter can be fabricated using two methods. The first way is the use of coating. Because the characteristic matrix for each layer is different for TE and TM waves for inclined angle of incidence according to Eq. (6-52) and Eq. (6-53). The overall characteristic matrix of the multiple layers is then different for the two polarizations. Therefore, the coating is usually polarization dependent for inclined angle of incidence. If the interface shown in Fig. 7-21 is properly coated, the TM wave will transmit and the TE wave will be reflected. The second way is the use of birefringence. If the cube is a birefringent material, the incident light will

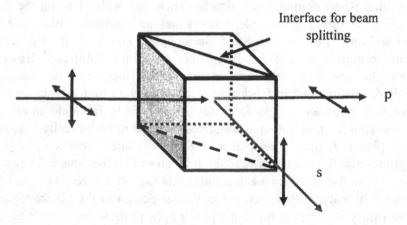

Figure 7-21. A polarization beam splitter separates different polarizations of light.

experience two different refractive indices, one for each polarization. The birefringent material is chosen such that the refractive index is smaller for horizontal polarization, so this polarization transmits the interface without reflection when it is incident at the Brewster angle, as explained in Chapter 1. For vertical polarization, the refractive index is larger, so its incident angle is larger than the critical angle. Thus total reflection will happen for this polarization. As a result, the two polarizations of light are separated.

Birefringence also enables us to use the polarization beam splitter in the waveguide type of combiners/splitters. According to Eqs. (6-67a) and (6-67b), the coupling coefficients are functions of the refractive index. For material of birefringence, the wave experiences two refractive indices. The one with the refractive index n_0 is the ordinary wave. The other with the refractive index n_e is the extraordinary wave. Their coupling coefficients are different. Therefore, it is possible to design the waveguide combiners/splitters so that the extraordinary wave has complete transfer of the optical power to the other waveguide, while the ordinary wave has the optical power transferred back to the original waveguide, or vice versa. In this way, the ordinary wave and the extraordinary wave are separated into the two outputs of the splitter. Because the ordinary wave and the extraordinary wave have perpendicular polarizations, this combiner/splitter works as a polarization beam splitter.

The polarization beam splitter has a unique advantage over the other combiners described above. It is able to completely combine two incident waves into one. This function is illustrated in Fig. 7-22. Because of the polarization selection on the transmission and reflection at the interface, the incident p-wave passes through the interface without reflection, while the s-wave is reflected at the interface without transmission. As the two beams are aligned properly, they combine to a single beam after the interface. Such combination of two beams to one beam does not violate the energy conservation. The analysis is as follows.

Assume that the power in the two input paths is the same, $P_1 = P_2 = \beta E^2$, where β is a proportional constant. However, the light from the two inputs has different polarizations. With the convention shown in Fig. 7-22, the p-wave is horizontally polarized, so its field is expressed as $\hat{x}Ee^{j\phi 1}$. The s-wave is vertically polarized, so its field is expressed as $\hat{y}Ee^{j\phi 2}$. The light at the output of the combiner then has the field $\vec{E}_{out}$ equal to the vector summation of the two fields from the two inputs.

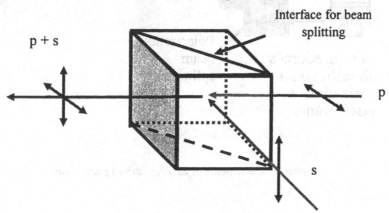

Figure 7-22. A polarization beam splitter combines two beams of different polarizations to one beam.

$$\vec{E}_{out} = \hat{x}Ee^{j\phi 1} + \hat{y}Ee^{j\phi 2} \qquad (7\text{-}26)$$

Eq. (7-26) differs from Eq.(7-1) by the polarization. Because the two input fields now have orthogonal polarizations, the total power becomes $P_{out} =$

$$\beta \left| Ee^{j\phi 1} \right|^2 + \beta \left| Ee^{j\phi 1} \right|^2 = 2\beta E^{\,2} = P_1 + P_2.$$ This result is independent of the

phases $\phi 1$ and $\phi 2$. Therefore, combination of two input beams to one is possible using the polarization beam splitter.

Due to the above characteristics, the polarization beam splitter is particularly useful for power combination. An example of its use is the addition of power from two pump sources of laser diodes in the EDFA, as illustrated in Fig. 7-23. The two pump sources are arranged in the way that their output laser beams have orthogonal polarizations. The laser beams are incident on the polarization beam splitter with the polarizations and the directions shown in Fig. 7-22.

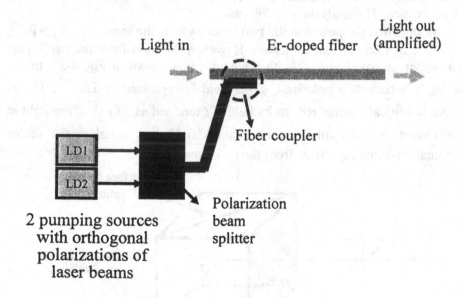

Figure 7-23. Schematic of EDFA with two pump sources.

3.   FIXED AND TUNABLE FILTERS

In a WDM optical communication system, each optical channel is characterized by an individual wavelength. Many channels are transmitted together in the system. It is sometimes necessary to select one particular channel from the system for special treatment or transmission. Because each channel is associated with a wavelength, we need the capability of optical selection in the system configuration. An optical filter is thus used for this purpose. It allows only one wavelength to pass through and blocks all other wavelengths, as illustrated in Fig. 7-24. In addition to blocking channels of other wavelengths, it can also filter out optical noise outside its bandwidth. There are fixed and tunable filters.

3.1 Fixed Filters

For fixed filters, the wavelength cannot be selected by an external control. Many methods can be used to fabricate fixed filters. As long as a device is wavelength selective, it will work as a fixed filter. Examples include but are not limited to:
- diffraction grating;
- thin-film filter;
- fiber Bragg grating;
- Mach-Zehnder interferometer.

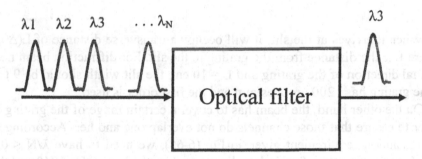

Figure 7-24. An optical filter selects optical signals at one wavelength ($\lambda 3$).

3.1.1 Optical Filter Based on Diffraction Grating

As explained in Chapter 6, the diffraction angle θ_r is a function the wavelength for $m \neq 0$, different wavelengths are separated by a grating and then propagate in different directions. As shown in Fig. 7-25, if we use a slit at the proper position, a particular wavelength is selected, while other wavelengths are blocked. The width of the slit depends on the bandwidth of the channel at the wavelength. If the channel spacing is 100 GHz and the channel of interest is centered at 1550 nm, then the spectral separation $\Delta\lambda$ in wavelength is 0.8 nm. From the principle of grating $\sin\theta_r = \sin\theta_i + m\dfrac{\lambda}{\Lambda}$, this spectral width will spread out spatially with an angle $\Delta\theta_r$.

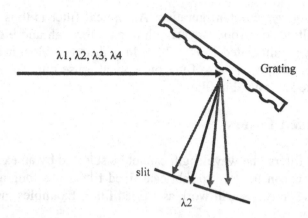

λ1, λ2, λ3, λ4

Grating

slit

λ2

Figure 7-25. Optical filter based on diffraction grating. The wavelength λ2 is selected behind the slit.

$$\Delta\theta_r = m\frac{\Delta\lambda}{\Lambda}\cos^{-1}\theta_r \qquad (7\text{-}27)$$

When it arrives at the slit, it will occupy a transverse distance of $L(\Delta\theta_r)$, where L is the distance from the grating to the slit. For diffraction beam near normal direction of the grating and L = 10 cm, the slit width should be 9 μm if the grating has 1200 grooves/mm and the first order is used.

On the other hand, the beam has to cover a certain range of the grating in order to ensure that those channels do not overlap one another. According to the resolution requirement given in Eq. (6-65), we need to have $\lambda/N < 0.8$ nm. Thus the number of periods to be covered by the beam $N > 1940$ and the range $N\Lambda > 1.6$ mm. A better situation is that the beam's size is more than twice the range required by the resolution limit of the grating, in this case > 3.2 mm.

Because the beam travels in free space, it should be packaged properly for the use in optical fiber communications. The package will be similar to the one for bulk-type combiners/splitters and will not be elaborated on here.

3.1.2   Thin-Film Filter

Similar to the bulk-type combiner/splitter, the thin-film is a glass or other transparent dielectric with a coating. Again, one of the interfaces is coated to exhibit the desired function, while the other interface is AR coated, as shown in Fig. 7-26 (a). Its function is illustrated in Fig. 7-26 (b). The multi-layer coatings shown have very strong reflection for the wavelength λ3, while other wavelengths transmit through the coatings without much loss. The incident angle may be 90° as shown in the figure or another angle, depending on the design. A typical spectral response is shown in Fig. 7-26 (c).

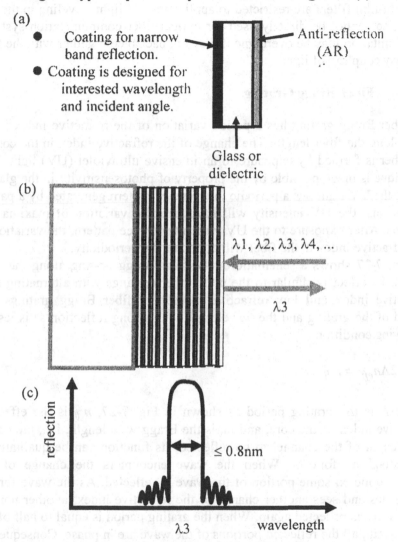

Figure 7-26. (a) Thin-film filter: one side is coated for reflection at the interested wavelength and the other side is AR-coated. (b) Function of the thin-film filter. (c) Spectral response of the thin-film filter.

Strong reflection at a desired center wavelength and bandwidth can be achieved using multi-layer coatings, as explained in Chapter 6. Usually pairs of quarter-wavelength layers with alternating high refractive index and low refractive index are used, just like those used for VCSEL. Therefore, the analysis in Sec 6.3 can be applied here. Typically the 3 dB bandwidth of 100 GHz (~ 0.8 nm) can be achieved using multi-layer coatings. To achieve 50

GHz bandwidth of reflection, the coating conditions are very stringent and cannot be easily achieved.

Thin-film filters are restricted to applications of light traveling in the free space, so cannot be directly used for optical fiber communication systems. This limitation may be overcome if they are packaged together with the fiber for easy coupling of light.

3.1.3   Fiber Bragg Grating

Fiber Bragg grating has a period variation of the refractive index in the core along the fiber length. The change of the refractive index in the core of the fiber is formed by exposure to an intensive ultraviolet (UV) light. This technique is made possible by the property of photosensitivity in the glass. [8] When the UV light has a periodic interference pattern generated by a pair of UV beams, the UV intensity will have periodic variation of maxima and minima. After exposure to the UV with interference pattern, the variation of the refractive index in the fiber will duplicate the periodicity. [9]

Fig. 7-27 shows a schematic of the fiber Bragg grating along the fiber length. Its effect is similar to the multi-layer coatings with alternating high refractive index and low refractive index. For fiber Bragg gratings, the period of the grating and the wavelength with strong reflection satisfies the following condition

$$2\Lambda n_{eff} = \lambda_B \tag{7-28}$$

where Λ is the grating period as shown in Fig. 7-27, n_{eff} is the effective refractive index of the core, and λ_B is the Bragg wavelength, i.e., the center wavelength of the channel to be reflected. Its function can be qualitatively understood as follows. When the wave encounters the change of the refractive index, some portion of the wave is reflected. As the wave further propagates and sees another change of the refractive index, another portion of the wave is reflected again. When the grating period is equal to half of the wavelength, all the reflected portions of the wave are in phase. Consequently, they have constructive interference and thus combine to form a strong reflected wave. The condition in Eq. (7-28) is the requirement to have constructive interference.

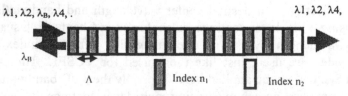

Figure 7-27. Schematic of index variation in the fiber Bragg grating.

Because the variation of the refractive index is similar to the multi-layer coatings, the analysis of multi-layer coatings can be used to understand its functional principle. The variation of the refractive index (n_1-n_2) is only in the order of 10^{-3}-10^{-4} though. To produce similar effects to those achieved with multi-layer coatings, many more periods of alternating refractive index are required. Fortunately the fiber Bragg grating is very long, typically around 10 mm, so the cumulative effect is also large. On the other hand, the light propagating in the fiber is a guiding mode, which deviates from the plane wave. Thus some theories have been developed to analyze the reflection of mode propagation in the fiber Bragg grating. [10, 11] The results actually do not deviate much from those obtained by the multi-layer coatings. Fig. 7-28 shows a typical spectral response of the fiber Bragg grating. It reflects a certain wavelength and so works as an optical filter.

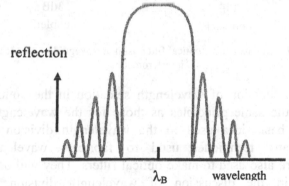

Figure 7-28. Spectral response on the reflection of a fiber Bragg grating.

3.1.4   Mach-Zehnder Interferometer and Other Types of Optical Filters

A Mach-Zehnder interferometer can also work as an optical filter. When the optical paths of the two arms in the Mach-Zehnder interferometer are properly designed, the light at the wavelength of interest will produce constructive interference at the output port, while light of other wavelengths will produce destructive interference. Thus only the desired wavelength is obtained at the output port. Fig. 7-29 illustrates an example of the wavelength selection using a waveguide-type Mach-Zehnder interferometer.

The 3 dB coupler1 splits the input light equally to path1 and path2. Path2 is longer than path1 for a length of ΔL. This length difference ΔL is designed in such a way that $n_{eff} \Delta L = m \lambda 1$, while another wavelength like $\lambda 2$ has $n_{eff}\Delta L = (n +1/2) \lambda 2$, where n and m are integers. When the waves in the two paths combine again at the 3 dB coupler2, the light at the wavelength $\lambda 1$

experiences constructive interference at port1, while other wavelengths experience destructive interference at the same port. Therefore, only the light at $\lambda 1$ is selected at port1. On the other hand, as we explained at Sec. 6.6, when the light in the Mach-Zehnder interferometer experiences destructive interference at one output, it should experience constructive interference at the other output. Thus, other wavelength $\lambda 2$ shown in Fig. 7-29 should experience constructive interference at port2. This Mach-Zehnder interferometer then works as a wavelength division multiplexer.

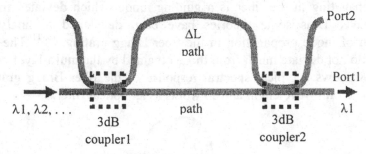

Figure 7-29. Schematic of an optical filter using a waveguide-type Mach-Zehnder interferometer.

In fact, the function of wavelength selection in the optical filter very often utilizes the same principles as those for the wavelength separation, which is the basic idea used in the wavelength division multiplexers. Therefore, many techniques used to fabricate wavelength division multiplexers are also used to make optical filters. They will be described in more detail in the discussion of wavelength division multiplexers/demultiplexers.

3.2   Tunable Filters

For tunable filters, the selected wavelength can be varied by some external control. There are also many methods to fabricate tunable filters. A tunable filter can be illustrated schematically by Fig. 7-30. They include
- Fabry-Perot tunable filter;
- tunable filter based on diffraction grating;
- tunable filter based on thin-film filter;
- tunable filter based on fiber Bragg grating;
- electro-optic tunable filter;
- acousto-optic tunable filter.

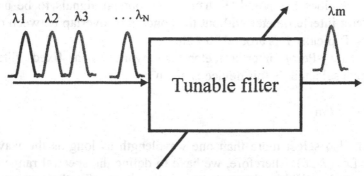

Figure 7-30. A schematic showing the concept of a tunable filter. The arrow indicates that the selected wavelength is variable.

3.2.1 Fabry-Perot Tunable Filter

In Sec. 6.2, we analyzed Fabry-Perot resonance. The cause of the resonance is the light bouncing forwards and backwards between the two parallel interfaces. If the two interfaces are replaced with two mirrors of high reflectivity, similar effects still occur. The device with two parallel mirrors is called Fabry-Perot interferometer or etalon. The light intensity or power transmitting the Fabry-Perot interferometer is characterized by the transfer function T_{FPF}, which is given by Eq. (6-21). The formula is rewritten here,

$$T_{FPF} = \frac{(1-R)^2}{(1-R)^2 + 4R\sin^2[(\omega - \omega_0)d/n]}$$, to facilitate discussion. It has

maximum transmission of power ω_0 that is determined by the condition: nd = m $(\lambda/2)$, where m is an integer, n is the refractive index of the material between the mirrors and d is the distance between the two mirrors. Therefore, the transfer function and the peaks periodically appear over the frequency domain or the wavelength domain. The spectral separation between the peaks of the transmission is called free spectral range (FSR). FSR = $\dfrac{\lambda^2}{2nd}$ (in wavelength) or FSR = $c/(2nd)$ (in frequency). The 3 dB bandwidth for each transmission line is $\dfrac{c}{2nd}\dfrac{1-R}{R^{1/2}}$, given in Eq.(6-22). The ratio of the FSR to the 3 dB bandwidth is called Finesse F.

$$F = \frac{\pi\sqrt{R}}{1-R} \qquad\qquad (7\text{-}29)$$

Finesse describes the possible channels of optical signals to be used in a Fabry-Perot interferometer without the concern of overlap between different channels. Typically F is around 20-100.

The Fabry-Perot interferometer can work as a fixed filter. The wavelength selected by this device is given by

$$\lambda = 2\,nd/\text{m} \qquad\qquad (7\text{-}30)$$

It can also select more than one wavelength as long as the wavelength satisfies Eq. (7-30). Therefore, we have to define the spectral range for it to work properly. Within this spectral range, if we fix the integer m, the wavelength λ and the distance d has one to one correspondence. If somehow we are able to change the distance d, the selected wavelength becomes tunable. Fig. 7-31 shows one configuration that has the capability. The two ends of the fibers are coated to exhibit high reflectivity, so the two fiber ends work as a Fabry-Perot filter. The two ends are separated with a gap d, which is adjusted using a piezo control as shown in Fig. 7-31. The peak wavelength of transmission in this configuration is therefore tuned by the piezo control.

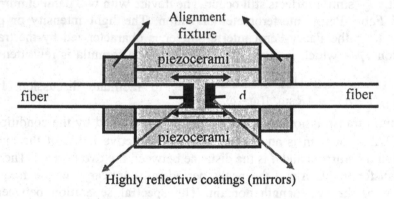

Figure 7-31. A Fabry-Perot tunable filter.

3.2.2   Tunable Filter Based on Diffraction Grating

The use of a diffraction grating as an optical filter has been explained before. The diffraction beam follows the principle of grating $\sin\theta_r = \sin\theta_i + m\dfrac{\lambda}{\Lambda}$. If the grating rotates, while the direction of incident beam remains the same, as shown in Fig.7-32, the incidence angle θ_i changes. As a result, the diffraction angle varies. Thus the directions for these diffracted wavelengths change accordingly. In Fig.7-32, the direction of the incident beam and the position of the slit are not changed when the grating rotates. In Fig.7-32 (a), the grating is at the position for the selection of

wavelength λ2. When the grating rotates to a new orientation shown in Fig.
7-32 (b), the incident angle changes, so all of the diffracted beams also

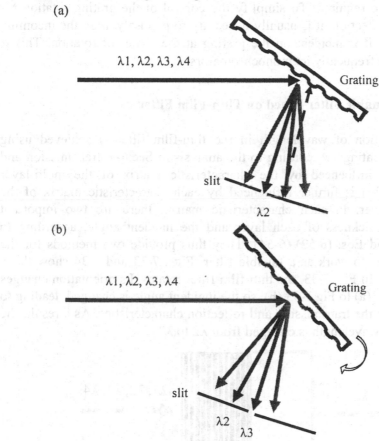

Figure 7-32. (a) The grating is at the position for the selection of wavelength λ2. (b) The grating is rotated to select another wavelength λ3.

change their directions. As a result, the beam that passes through the slit
changes to another wavelength λ3. The rotation angle of the grating to make
the selection from wavelength λ2 to λ3 is equal to Δθ, given by

$$\Delta\theta = m \frac{|\lambda 2 - \lambda 3|}{\Lambda} \cos^{-1}\theta \qquad (7\text{-}31)$$

When the grating rotates, ideally it will just change the directions of
diffracted beams. However, if the center of rotation is not at the point where
the light is incident on the grating, the diffracted beams will not only change
directions, but also experience lateral translation. Then even though the
grating is rotated with an angle given by Eq. (7-31), the beam of the

interested wavelength will not pass through the slit. Some tedious calculations on the exact angle of rotation to compensate for the lateral translation are required. To simplify the control of the grating rotation for wavelength selection, it is usually necessary to precisely align the incoming beam so that it is incident on the grating at the center of rotation. This is actually done frequently in monochromators.

3.2.3   Tunable Filter Based on Thin-Film Filter

The selection of wavelength in the thin-film filter is achieved using multi-layer coatings. According to the analysis in Sec.6-3, transmission and reflection is influenced by the characteristic matrix of the multi-layer coatings, which is further influenced by each characteristic matrix of the individual layer. In each characteristic matrix, there are two important factors: the thickness of each layer and the incident angle, according to Eq.(6-43), and Eqs. (6-52)-(6-54). They thus provide two methods for the thin-film filter to work as a tunable filter. Figs. 7-33 and 7-34 show those two methods. In Fig. 7-33, the thin-film filter rotates. Its orientation changes from Fig. 7-33(a) to Fig.7-33 (b), so the incident angle is changed, leading to the change of the transmission and reflection characteristics. As a result, the transmission wavelength is changed from $\lambda 2$ to $\lambda 3$.

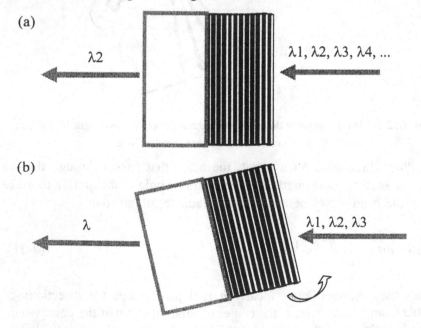

Figure 7-33. (a) The thin-film filter is at the position for the selection of wavelength $\lambda 2$. (b) The thin-film filter is rotated to select another wavelength $\lambda 3$.

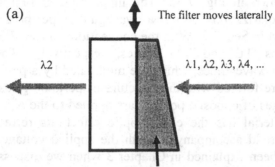

(a) The filter moves laterally

$\lambda 2$ $\lambda 1, \lambda 2, \lambda 3, \lambda 4, ...$

The thickness of the thin films is gradually changed.

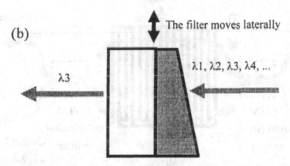

(b) The filter moves laterally

$\lambda 3$ $\lambda 1, \lambda 2, \lambda 3, \lambda 4, ...$

Figure 7-34. (a) The thin-film filter is at the position for the selection of wavelength $\lambda 2$. (b) The thin-film filter moves laterally to select another wavelength $\lambda 3$.

In Fig. 7-34, the multi- layer coatings are made in such a way that the thickness gradually varies along one direction. When the filter moves laterally from the position shown in Fig. 7-34 (a) to a second position shown in Fig. 7-34 (b), the transmission and reflection characteristics are different due to the change of film thickness. Thus the selected wavelength changes accordingly.

3.2.4 Tunable Filter Based on Fiber Bragg Grating

According to the condition for fiber Bragg gratings, the reflected light with the Bragg wavelength λ_B and the grating period Λ satisfy the equation: $2\Lambda n_{eff} = \lambda_B$. If the period is changed, the Bragg wavelength also changes. The grating period can be varied by stretching the fiber or heating the fiber to induce thermal expansion. The reflected wavelength thus varies with the external control.

3.2.5 Electro-Optic Tunable Filter

The electro-optic tunable filter utilizes material with birefringence and electro-optic effects. [12] It is usually made in the waveguide-type. A schematic is shown in Fig. 7-35. The input light is first split into two perpendicular polarizations using a waveguide-type polarization beam splitter, explained in Sec. 7-2. With the conventions in Fig. 7-35, we call the two polarizations TE and TM waves, respectively. They propagate separately in two waveguides, which are modulated by a pair of electrodes. The electrodes are formed with the structure of periodic grids, as shown in Fig. 7-35. Voltages of opposite polarity are applied to the pair of electrodes. Because the material has the electro-optic effect, its refractive index is changed by the field accompanying with the applied voltage. The electro-optic effect has been explained in Chapter 3 when we discussed the Mach-Zehnder modulator. To facilitate explanation, we rewrite Eq. (3-10) here.

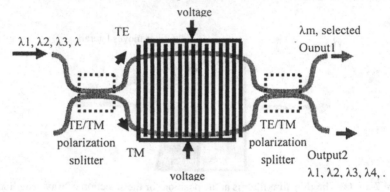

Figure 7-35. An electro-optic tunable filter.

$$(\frac{1}{n_x^2} + r_{1k}E_k)x^2 + (\frac{1}{n_y^2} + r_{2k}E_k)y^2 + (\frac{1}{n_z^2} + r_{3k}E_k)z^2 + 2yzr_{4k}E_k + 2zxr_{5k}E_k + 2xyr_{6k}E_k = 1$$

The above equation describes two points. First, the refractive index changes with the applied field. Second, the original principal axis of the index ellipsoid rotate to new axis. Therefore, the original TE wave is no longer the eigenmode of the material of birefringence when the external field is applied. It then decomposes into two modes of perpendicular polarizations according to the new principal axis of the index ellipsoid. The two decomposed modes also experience different refractive indices, so their phases vary as they propagate. As a result, the superposition of the two modes has a resultant field with polarization rotating along the propagation distance. Similarly, the TM wave in the other waveguide has its polarization rotating along the propagation distance. The electrodes with periodic grids induce periodic variation of the refractive index that could lead to rotation of polarization. They are designed in such a way that the wavelength of interest will have the TE wave change to TM polarization and the TM wave change to TE polarization when they are going to combine at the second polarization The change of polarization for 90° only happens for the phase-matched wavelength. For other wavelengths that are not phase matched, their polarizations remain the same.

When the waves in the two waveguides combine at the second polarization beam splitter, the wavelength with polarization changed by 90° is delivered to output1. The other wavelengths with no change of polarization are delivered to output2. Because the phase-matched condition varies with the strength of the applied field, i.e., the applied voltage, the wavelength selected at ouput1 varies with the applied voltage, thus making it work as a tunable filter.

Compared to other types of tunable filters, the electro-optic tunable filter is more complicated to fabricate. However, it has the advantage that its tuning speed is very fast (~ ns). The 3 dB bandwidth of the selected channel is about 1 nm, which can be reduced by increasing the periodic number of the electrode grids at the expense of increased attenuation. It is unfortunate that the tuning range is only about 10 nm. In comparison, the tuning range of acousto-optic tunable filters is much larger although the working principle is similar. The acousto-optic tunable filter will be discussed next.

3.2.6   Acousto-Optic Tunable Filter

A schematic of an acousto-optic tunable filter is shown in Fig. 7-36. Its configuration is similar to the electro-optic tunable filter. A material with birefringence and acousto-optic effect is used. [12] It is also made in the waveguide-type. Again, the input light is first split into a TE wave and a TM wave that propagate separately in the two waveguides. In contrast to the electro-optic tunable filter, the rotation of polarization is now caused by the applied acoustic wave. Therefore, there are no electrodes with the structure of periodic grids to create the periodic variation of the refractive index using electro-optic effect. Instead, the periodic variation of the refractive index is created by the surface acoustic wave (SAW). A transducer is thus fabricated on the surface. When RF signals are applied to the transducer, a surface acoustic wave is generated. As long as the periodic variation of the refractive index is created, the phase-matched wavelength has its polarizations rotated 90°. Thus the TE wave in the top waveguide becomes TM-polarized and the TM wave in the bottom waveguide becomes TE-polarized. When they combine at the second polarization beam splitter, the change of polarization causes them to be delivered to output1. In contrast, other wavelengths have no change in their polarization, so they are delivered to output2.

The acousto-optic effect is similar to the electro-optic effect. Because the acoustic wave causes the vibration of atoms in the materials, the refractive index varies with the amplitude of the acoustic wave. In addition, the crystal has a certain lattice structure, the vibration of atoms then does not completely follow the direction of strain fields in the acoustic wave. In general, the material with the acousto-optic effect has its index changed according to the following formula for the index ellipsoid. [13]

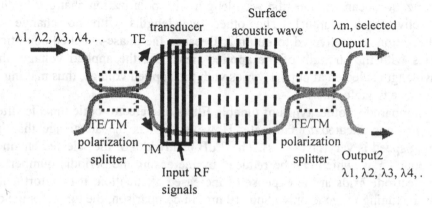

Figure 7-36. An acousto-optic tunable filter.

$$x^2(\frac{1}{n_x^2} + p_{11}S_1 + p_{12}S_2 + p_{13}S_3 + p_{14}S_4 + p_{15}S_5 + p_{16}S_6) +$$

$$y^2(\frac{1}{n_y^2} + p_{21}S_1 + p_{22}S_2 + p_{23}S_3 + p_{24}S_4 + p_{25}S_5 + p_{26}S_6) +$$

$$z^2(\frac{1}{n_z^2} + p_{31}S_1 + p_{32}S_2 + p_{33}S_3 + p_{34}S_4 + p_{35}S_5 + p_{36}S_6) +$$

$$2yz(p_{41}S_1 + p_{42}S_2 + p_{43}S_3 + p_{44}S_4 + p_{45}S_5 + p_{46}S_6) +$$

$$2zx(p_{51}S_1 + p_{52}S_2 + p_{53}S_3 + p_{54}S_4 + p_{55}S_5 + p_{56}S_6) +$$

$$2xy(p_{61}S_1 + p_{62}S_2 + p_{63}S_3 + p_{64}S_4 + p_{65}S_5 + p_{66}S_6) = 1 \quad (7\text{-}32)$$

where the subscripts have the same meanings as those defined for the electro-optic effect. 1=(xx), 2=(yy), 3=(zz), 4=(yz)=(zy), 5=(xz)=(zx), 6=(xy)=(yx). The first index represents the direction of the strain field, while the second index indicates the propagation direction of the acoustic wave.

For example, if the acoustic wave is propagating along the z-axis and its vibration has components for all three directions, we can express it as

$$\vec{u}(z,t) = (A_1\hat{x} + A_2\hat{y} + A_3\hat{z})\cos(\Omega t - Kz) \qquad (7\text{-}33)$$

The strain field is then given by

$$\vec{S}(z,t) = K(A_1\hat{x} + A_2\hat{y} + A_3\hat{z})\sin(\Omega t - Kz) \qquad (7\text{-}34)$$

Therefore, we have

$$\begin{cases} S_5 = KA_1, & (xz) \\ S_4 = KA_2, & (yz) \\ S_3 = KA_3, & (zz) \end{cases} \qquad (7\text{-}35)$$

where A_i is the amplitude of the acoustic wave along that direction. K and Ω are the wave number and the frequency of the acoustic wave, respectively.

p_{ij} is the strain-optic coefficient, usually defined in the principal coordinate system of the crystal. From Eq. (7-32), we realize that the existence of the strain fields has two effects. First, the refractive index changes with the applied field. Second, the original principal axis of the index ellipsoid rotate to new axis. Therefore, the acousto-optic effect could lead to same polarization rotation as we discussed for the electro-optic effect.

Because the periodic variation of refractive index is now induced by an acoustic wave, this period corresponds to the wavelength of the acoustic wave. By changing the wavelength of the acoustic wave, the period varies accordingly. This changes the phase-matched wavelength. The flexibility of selection on the acoustic wavelength makes phase-match possible for a much broader spectral range. The tuning range can thus achieved for the spectral range from 1.3 μm to 1.6 μm. [14] The 3 dB bandwidth of the selected channel is similar to the electro-optic tunable filter, which is about 1 nm. It can also be reduced by extending the range with SAW at the expense of increased attenuation. [15]

The tuning speed is limited to approximately 1μs. For the acousto-optic effect to work for the phase-matched wavelength, the SAW has to fill up the interaction region. The tuning time is thus limited by the slow speed of the acoustic wave.

3.3   Specifications of Filters

The optical filters, including fixed filters and tunable filters, usually have the following specifications:
- center frequency or center wavelength: comply with ITU-T grid;
- pass bandwidth, defined as the width of a filter's transmission curve at 0.5 dB level of the maximum transmission peak, also called channel width or passband width;
- stop bandwidth, defined as the linewidth at the 20-dB level of the maximum transmission peak;
- ripple, defined as the peak-to-peak variation of the transmission within the channel width;
- loss, or insertion loss;
- polarization-dependent loss;
- sidelobe suppression ratio (SSR), defined as the ratio of maximum power of the first side peak to the power of the main peak (in dB);
- resolution, defined as the minimum shift of a wavelength that a filter can detect.

Tunable filter specifications include the following additional items:
- dynamic (tuning) range, defined as the range that the filter can handle, also called wavelength tunability;
- number of resolvable channels, defined as the ratio of the dynamic range to the minimum channel spacing;
- tuning speed, defined as time needed to tune a filter to a specific wavelength.

4.   ISOLATOR

An isolator allows the forward light to pass through with minimal loss and blocks backward light with maximum loss. In an optical communication system, reflection of light may occur from any components like connectors, mechanical splices, passive components, receivers, and so on. Also, the amplified spontaneous emission (ASE) from an optical amplifier could travel toward the direction opposite to the signal flow. The optical power flowing back to the active components like lasers or optical amplifiers could cause undesired effects. Therefore, it is important to prevent the light from going backward to the active components. Isolators are then required in the optical communication system to guarantee the direction of signal transmission. The function of an isolator is illustrated in Fig. 7-37.

4.1   Polarization Dependent Isolators

In practice, an isolator consists of two polarizers and one Faraday rotator that are arranged in the order shown in Fig. 7-38. The Faraday rotator is placed between the two polarizers. The output polarizer is rotated 45° from the input polarizer. The forward light, indicated by red lines, first passes through the input polarizer that causes the unpolarized light to be vertically polarized. Afterwards, it enters the Faraday rotator that is under a magnetic field pointing to right direction. The Faraday rotator then rotates the polarization to 45° from the original polarization. After leaving the Faraday rotator, the forward light passes through the output polarizer, which has the selection of polarization 45° from the input polarizer.

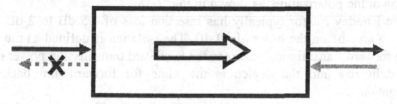

Figure 7-37. The function of an isolator: forward light (represented as red arrow) passes through, while backward light (represented as orange arrow) is blocked. The large arrow sign in the center indicates the direction along which it allows light to pass through.

For the backward light, the output polarizer causes its polarization to be 45° from the vertical direction before it enters the Faraday rotator. When it passes through the Faraday rotator, its polarization further rotates to the horizontal direction. The rotation direction is the same as that for input light. Because the input polarizer only selects the vertical polarization, the backward light is blocked.

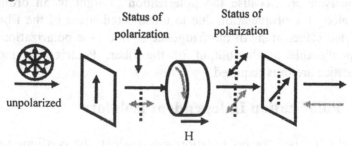

Figure 7-38. The configuration of an isolator.

The Farady rotator is made of material that has properties similar to optical activity. When the Faraday rotator is placed in a static magnetic field, the light traveling in it has its polarization rotating along the propagating distance, as we explained in Sec. 6.8. This phenomenon is called Faraday effect. The angle of rotation ϕ is proportional to the distance that light travels.

$$\phi = \rho HL \tag{7-36}$$

where H is the magnetic field in the unit of A/m, L is the length of the Faraday rotator length in the unit of meter, and ρ is the Verdet constant.

Although the polarization rotation due to the Faraday effect and the optical activity are similar, there is still a major difference. The polarization rotation in the Faraday rotator is nonreciprocal. The rotation direction is determined by the direction of the magnetic field instead of the direction of the wave vector. Thus the reverse propagation of light does not reverse the rotation of the polarization, as shown in Fig. 7-38.

The Faraday rotator typically has insertion loss of 0.5 dB to 2 dB. The isolation may be on the order of 40 dB. The isolation is defined as the ratio of the forward transmission power to the backward transmission power if the light delivered into the device is the same for forward and backward propagation.

The above isolator is polarization dependent. If the input light is polarized initially, the transmission power at the output power will depend on the polarization. For example, if the input light is vertically polarized, the input polarizer will allow the light to pass through completely. It then follows the polarization rotation and transmits through the output polarizer. We thus obtain the output light with the power equal to the input power minus the insertion loss. In contrast, if the input light is horizontally polarized, the input polarizer will block this light. As a result, we obtain no output light. The polarization dependence makes it inconvenient to use the isolator in the optical communication system when the polarizer is used as an in-line isolator. Because the polarization of light in an ordinary fiber changes along the propagation due to unexpected stress of the fiber and the accumulated effect of small birefringence in fiber, the polarization direction is not predictable at the output of the fiber. Polarization independent characteristics are thus required.

4.2   Polarization Independent Isolators

To make the isolator polarization independent, the configuration shown in Fig. 7-39 is used. [16] The input light is first split into two perpendicular polarizations using a polarization beam splitter. The beam in path1 has its polarization changed 90° by a half-wave ($\lambda/2$) plate and so becomes the same as the other beam. Both beams then enter the Faraday rotator that rotates their polarization 45°. The beam in path2 passes another half-wave ($\lambda/2$)

plate, so its polarization is perpendicular to the beam in path1. Both beams then combine again with another polarization beam splitter. Note that the polarization beam splitter at the output may be placed in a different orientation from the polarization beam splitter at the input in order to have proper combination of the two perpendicularly polarized beams.

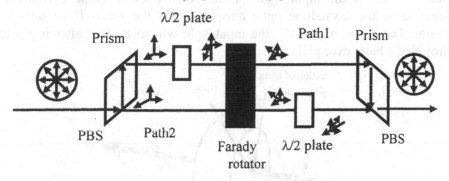

Figure 7-39. The configuration of a polarization-independent isolator.

For the backward light, its polarization is the same as the forward light before it enters the Faraday rotator. After the Faraday rotator, its polarization becomes perpendicular to the forward light for both paths even for the beam after passing through the half-wave ($\lambda/2$) plate in path1. When the two beams of the backward light enter the polarization beam splitter at the left hand side, they do not trace the same route as the forward light because of the different polarizations. Thus they do not go back to the initial path of the forward light.

The half-wave plate is made of materials with birefringence. In sec. 6-7, we had explained that the wave in birefringent material sees two refractive indices. One is the ordinary wave that experiences the refractive index n_0. The other is the extraordinary wave that sees with the refractive index n_e. Their polarizations are perpendicular to one another. When the birefringent crystal is cut with the optic axis parallel to the cutting plane, the two waves incident normally on the cutting plane of the crystal see the largest difference of refractive indices. After they pass through the crystal, they have different phase delay. The phase delays for the ordinary wave and the extraordinary wave are $\delta_0 = \dfrac{2\pi}{\lambda} n_0 h$ and the $\delta_e = \dfrac{2\pi}{\lambda} n_e h$, respectively, where h is the thickness of the cut crystal, i.e., the distance that the light passes through. Therefore, they have a phase difference

$$\delta = \delta_e - \delta_0 = \frac{2\pi}{\lambda}(n_e - n_0)h \qquad (7\text{-}37)$$

When the phase delay $\delta = \dfrac{\pi}{2}, \dfrac{3\pi}{2}$, ..., it is called quarter-wave ($\lambda/4$) plate. If the phase delay $\delta = \pi, 3\pi$, ..., it is called half-wave ($\lambda/2$) plate. The half-wave plate has the effect shown in Fig. 7-40. The polarization of light is rotated by 2θ if the input light has its polarization at angle θ from the direction of the crystalline optic axis, which is the z-axis for a uniaxial crystal. Therefore, if $\theta = 45°$, the input light will rotate $90°$ after it passes through the half-wave plate.

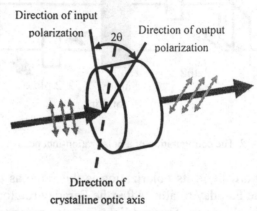

Figure 7-40. Effect of a half-wave plate on the polarization of the input light.

5.   CIRCULATOR

A circulator is a multi-port device. The function of a circulator with four ports is illustrated in Fig. 7-41. Input power from port 1 is delivered to port 2, input power from port 2 is delivered to port 3, input power from port 3 is

Figure 7-41. A schematic showing the function of a circulator.

delivered to port 4, and input power from port 4 is delivered to port 1. Ideally, no power is delivered from one port to another at the order opposite to the above sequence.

The working principle is demonstrated in Fig. 7-41 for a four-port circulator. [17] This circulator consists of two polarization beam splitters, two prisms, one Faraday rotator, and one half-wave plate arranged in the order shown in Fig. 7-42. In Fig. 7-42 (a), input light comes from port 1. It is split into the vertical polarization and the horizontal polarization by the first polarization beam splitter as shown in the figure. The top beam is reflected by the prism, so it travels along the same direction as the bottom beam. After they pass through the Faraday rotator, their polarizations are both rotated 45°, but still remain perpendicular to one another. Afterwards, they pass through the half-wave plate, which is arranged to have the direction of the optic axis 22.5° from the horizontal direction. Therefore, the polarization of light in the top path rotates 45° and becomes horizontal polarization. On the other hand, the polarization of light in the bottom path is 67.5° from the direction of the optic axis before it enters the half-wave plate. After the half-wave plate, the polarization of this light rotates 135° and becomes vertical polarization. Therefore, the two output waves have their polarizations perpendicular to one another and perpendicular to their original polarizations. The bottom beam is further reflected by a prism so that it can combine with the top beam in the polarization beam splitter. When they combine in the second polarization beam splitter, they go out at port 2.

When input light comes from port 2. It is split again to the vertical polarization and the horizontal polarization by the second polarization beam splitter as shown in Fig. 7-42 (b). The light in the top path has horizontal polarization, which is 22.5° from the direction of the optic axis of the half-wave plate. After it passes through the half-wave plate, its polarization rotates 45° from the horizontal direction, as shown in the figure. The light in the bottom path has vertical polarization, which is 67.5° from the direction of the optic axis before it enters the half-wave plate. After the half-wave plate, the polarization of this light rotates 135° and also becomes 45° from the horizontal direction, but is perpendicular to the polarization of the light in the top path, as shown in the figure. Then they pass through the Faraday rotator that rotates both polarizations by 45°. The light in the top path returns to horizontal polarization and the light in the bottom path returns to the vertical polarization. When they combine in the first polarization beam splitter, they go out at port 3.

The light input from port 3 or port 4 can be traced similarly. They will go out to the port according to the same order.

There are several important specifications for circulators. First, the circulator can provide isolation for the undesired direction, for example, from port 2 to port 1 in the above case. The isolation may be as high as 70 dB. Second, the insertion loss may be as low as 0.6 dB. Third, the circulator has to be polarization independent. The polarization dependent loss (PDL) may be as low as 0.05 dB. Fourth, the surface of a circulator should have very small reflectivity. The return loss may be more than 50 dB. Fifth, because the circulator operates with orthogonal polarizations, it has polarization mode dispersion (PMD). PMD is defined as the time difference between signals of orthogonal polarizations travelling in a device. Typical values of PMD in circulators are about 0.1ps, but could be as small as 0.01 ps. Sixth, the spectral range for operation is about 40 nm, entered at either 1310 nm or 1550 nm.

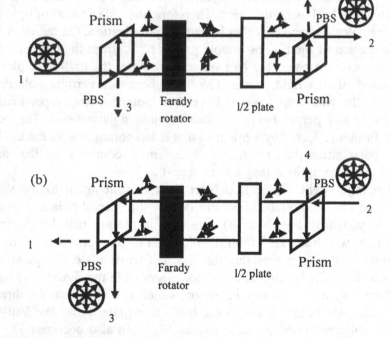

Figure 7-42. The configuration of a 4-port circulator. (a) Input light is form port 1. It goes to port 2. (b) Input light is form port 2. It goes to port 3.

6.   ATTENUATOR

Attenuators are used to reduce the light power in a controlled way. They are used for the following applications:
- prevent the saturation of a receiver due to a strong light intensity;
- equalize the power of different channels in the WDM system;

− test and evaluate components, modules, and systems;
− laboratory experiments.

Several methods have been used to fabricate attenuators. They will be described in the following.

6.1 Attenuators Using Absorbing Materials

Fig. 7-43 shows the method of using an absorbing material to reduce the optical power. The absorbing material is inserted between two fibers. If the absorption is fixed, the amount of attenuation is fixed, so this device is a fixed attenuator. If the absorption is variable, the device is a variable attenuator. The variable absorption could be achieved using the thermal-optical, electro-optical, or magneto-optical effects. [18]

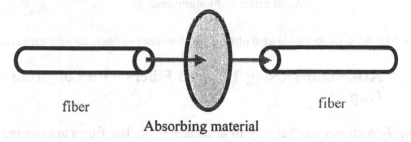

fiber   fiber

Absorbing material

Figure 7-43. An attenuator using the absorbing material for light attenuation.

6.2 Attenuators Using a Partially Reflective Mirror

An attenuator using a partially reflective mirror is shown in Fig. 7-44. This mirror is not normal to the light path in order to prevent light from going back to the light sources of the system. If the orientation of the mirror is fixed, it is a fixed attenuator. If the orientation is adjustable with some external control, it is a variable attenuator.

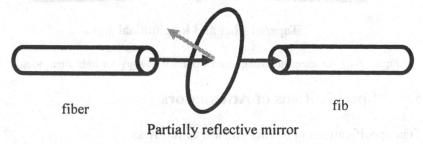

fiber   fib

Partially reflective mirror

Figure 7-44. An attenuator using a partially reflective mirror for light attenuation.

6.3   Attenuators Using Axial Offset or Longitudinal Gap of fibers

The third way to fabricate optical attenuators is to slightly misalign two fibers or to create a gap between two fibers, as shown in Fig.7-45. If the axial offset and the gap are fixed, it is a fixed attenuator. If the axial offset and the gap adjustable with some external control, it is a variable attenuator.

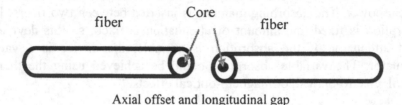

Axial offset and longitudinal gap

Figure 7-45. An attenuator using axial offset or gap between two fibers for light attenuation.

6.4   Attenuators Using Tapered Fibers with Longitudinal Gap

Fig. 7-46 shows another way to attenuate light. The fibers are tapered to have a small aperture. Light emerging out of the fiber then diverges quickly. As a result, the amount of light captured by the second fiber is small. The two fibers are usually precisely aligned to have no axial offset. The attenuation is a function of the gap between the two fibers. Again, if the gap is fixed, it is a fixed attenuator. If the gap is adjustable, it is a variable attenuator.

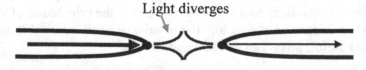

Tapered fiber and longitudinal gap

Figure 7-46. An attenuator using tapered fibers with a gap for light attenuation.

6.5   Specifications of Attenuators

The specifications of attenuators are as follows:
- for fixed type, the attenuation is usually predetermined, for example, 5 dB, 10 dB, and so on;
- the tolerance is usually ≤ 10 %;

- for variable type, the attenuation is typically variable from 1.5 dB to 60dB, with some up to 100 dB;
- the variation resolution of the variable attenuation could be set to 0.5 dB, o.1 dB, or even 0.01 dB;
- the input light power is limited to 20 dBm - 25 dBm;
- return loss typically larger than 55 dB;
- spectral range of operation is about 50 nm, expressed as, e.g., 1550 nm ± 25 nm.

REFERENCES

1. Kashima, Norio, *Passive Optical Components for Optical Fiber Transmission.* Artech House, 1995.
2. Shani, Y., Henry, C. H., Kistler, R. C., Kazarinov, R. F., and Orlowsky, K. J., Integrated optic adiabatic devices on silicon. IEEE Journal of Quantum Electronics 1991; 27: 556-566.
3. Jinguji, K., Takato, N., Sugita, A., and Kawachi, M., Mach-Zehnder interferometer type optical waveguide coupler with wavelength-flattened coupling ratio. Electronics Letters 1990; 26: 1326-1327.
4. Hanaizumi, O., Miyagi, M., Minakata, M., and Kawakami, S., Antenna coupled Y junction in 3-dimensional dielectric waveguide. European Conference on Optical Communication (ECOC), 1985, p.179.
5. Mestdagh, Denis J. G., Fundamentals of Multiaccess Optical Fiber Networks. Artech House, 1995.
6. Arkwright, J.W. and Mortimore, D. B., 7 x 7 monolithic single-mode star coupler. Electronics Letters 1990; 26: 1534-1536.
7. Takahashi, H., Okamoto, K., and Ohmori, Y., Integrated-optic 1 x 128 power splitter with multichannel waveguide. IEEE Photonics Technology Letters 1993; 5: 58-60.
8. Hill, K. O., Fujii, Y., Johnson, D. C., and Kawasaki, B. S., Photosensitivity in optical fiber waveguides: application to reflection filter fabrication. Applied Physics Letters 1978; 32: 647-649.
9. Hill, K. O and Meltz, G., Fiber Bragg grating technology: fundamentals and overview. IEEE Journal of Lightwave Technology 1997; 15: 1263-1276.
10. Erdogan, T., Fiber grating spectra. IEEE Journal of Lightwave Technology 1997; 15: 1277-1294.
11. Peral, E. and Capmany, J., Generalized Bloch wave analysis for fiber and waveguide gratings. IEEE Journal of Lightwave Technology 1997; 15: 1294-1302.
12. Green, Paul E. Jr., *Fiber Optic Networks.* Prentice Hall, 1993.
13. Yariv, Amnon and Yeh, Pochi, *Optical Waves in Crystals.* John Wiley & Sons, 1984.
14. Cheung, K. W., Choy, M. M., and Kobrinski, H., Electronic wavelength tuning using acoustooptic tunable filter with broad continuous tuning range and narrow channel spacing. IEEE Photon Technology Letters 1989; 1: 38-40.
15. Herrman, H., Müller-Reich, P., Reimann, V., Ricken, R., Seibert, H., and Sohler, W, Integrated optical TE- and TM-pass, acoustically tunable, double-stage wavelength filters in LINbO$_3$, Electronics Letters 1992; 28: 642-644.
16. Shiraishi, K., New configuration of polarization-independent isolator using a polarization-dependent one. Electronics Letters 1991; 27: 302-303.

17. Iwamura, H., Iwasaki, H., Kubodera, K., Torii, Y., and Noda, J., Simple-polarization-independent optical circulator for optical transmission systems. Electronics Letters 1979; 15: 830-831.
18. Backer, M. R., Electronic variable optical attenuators advance optical networking. Lightwave 1999; February: 122-124.

Chapter 8

PASSIVE COMPONENTS (II)

1. INTRODUCTION

With an understanding of the basic passive components provided by previous chapters, we can now study the more complex passive components. They are often formed from the basic passive components. Those complex passive components are sometimes composed of several simple components, but can also be fabricated using monolithic integration methods. Multiple schemes can also be used to fabricate each type of those components to achieve a similar function. Once again, we will highlight the principles applied and basic components used to form the complex components as well as how the functions are achieved as we go through this chapter.

The components to be discussed in this chapter are:
- wavelength division multiplexer/demultiplexer;
- optical add/drop multiplexer/demultiplexer;
- optical switch;
- optical crossconnect;
- wavelength router;
- wavelength converter.

2. WAVELENGTH DIVISION MULTIPLEXER/ DEMULTIPLEXER (WDM)

In current optical communication systems, many optical channels of signals transmit simultaneously in an optical fiber. Each channel has its own center wavelength and bandwidth. To put them together in one single fiber, we need a device called wavelength division multiplexer, also called a WDM MUX. At the receiving end, we have to separate those different optical channels from a single fiber in order to send them to their final destinations. We accomplish this with a device called a wavelength division demultiplexer, also known as WDM DEMUX. Optical communication systems that have many optical channels transmitted in a single fiber is thus called wavelength division multiplexing (WDM) system.

Both WDM MUX and WDM DEMUX are sometimes simply called WDM. Their functions are illustrated in Fig. 8-1. In fact, a MUX and DEMUX use the same operating principles. In many cases, they are the same devices simply used in the opposite way. The details will be discussed later. In addition, the functions of wavelength selection and combination in the WDM apply most of the same principles as those for optical filters. Therefore, many schemes used to fabricate optical filters can be used to fabricate wavelength division multiplexers/demultiplexers.

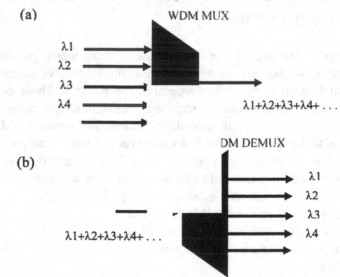

Figure 8-1. (a) A WDM MUX combines signals at different wavelengths. (b) A WDM DEMUX separates signals at different wavelengths.

Fig. 8-2 shows the characteristics of the optical channels delivered in an optical fiber or other device in the WDM system. The center wavelengths of those channels are separated by 0.8 nm, which corresponds to a frequency separation of 100 GHz. In fact, the frequency separation is fixed at 100 GHz, while the wavelength separation varies slightly because the center wavelength is different. The wavelength separation $(\Delta\lambda)$ and the frequency separation (Δf) are related by

$$|\Delta\lambda| = \frac{\lambda^2}{c}\Delta f \qquad (8\text{-}1)$$

where c is the speed of light and λ is the center wavelength of that channel. The neighboring channels overlap at the power at least 20 dB below their peak values in order to guarantee negligible crosstalk between them. Therefore, WDM MUX and DEMUX should satisfy the above requirements when they combine or separate those optical channels. [1]

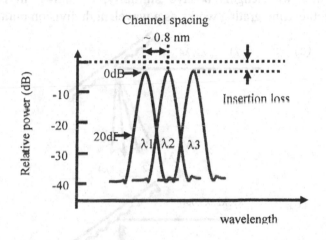

Figure 8-2. Characteristics of the optical channels delivered in the WDM system.

2.1 Grating-Type WDM

According to the principle of grating, $\sin\theta_r = \sin\theta_i + m\dfrac{\lambda}{\Lambda}$, the diffraction angle θ_r varies with the wavelength for $m \neq 0$. Therefore, different wavelengths are separated by the grating and propagate toward different directions. This gives the function for wavelength division demultiplexing, as shown in Fig. 8-2 (a). For example, the diffraction angle for the wavelength $\lambda 1$ is

$$\sin\theta_{r1} = \sin\theta_i + \frac{\lambda 1}{\Lambda} \qquad (8\text{-}2)$$

for the order $m = 1$.

On the other hand, when the light of wavelength $\lambda 1$ is incident on the grating from the diffraction direction at the above angle, it will propagate toward the direction defined by the original incident angle. That is, we have the new incident angle $\theta_i' = \theta_{r1}$ and the new diffraction angle $\theta_{r1}' = \theta_i$. Substituting them into Eq. (8-2), we obtain

$$\sin\theta_{r1}' = \sin\theta_i' - \frac{\lambda 1}{\Lambda} \qquad\qquad (8\text{-}3)$$

Eq. (8-3) means that the new diffraction beam is obtained for the order m = -1. The other wavelengths behave similarly, as shown in Fig. 8-2 (b). Therefore, the same grating works as a wavelength division multiplexer.

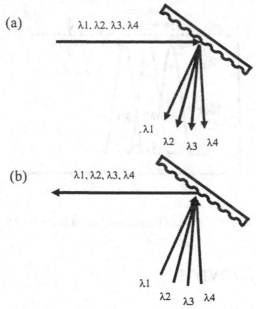

Figure 8-3. Use of grating for (a) wavelength division demultiplexing (b) multiplexing.

Using diffraction for multiplexing and demultiplexing, the beam travels in the free space. Thus it should be packaged properly for the use in fiber optic communications. In practice, the diffraction grating is often used together with a glass wedge and a GRIN-rod lens so that light can be easily collected by the fiber. Fig. 8-4 schematically shows the combination of those components and the light paths of different wavelengths for its use as a WDM MUX or WDM DEMUX. The graded index (GRIN) rod lens behaves like the fiber. The light can be confined in the GRIN-rod lens. Its center refractive index is the largest and decreases gradually with the transverse radius. The light path therein follows a sinusoidal curve. The variation over one period is called one pitch. The GRIN rod lens is usually made with a length slightly less than one quarter pitch. It causes the light path to follow the curve shown in Fig. 8-4, so it functions similar to a spherical lens.

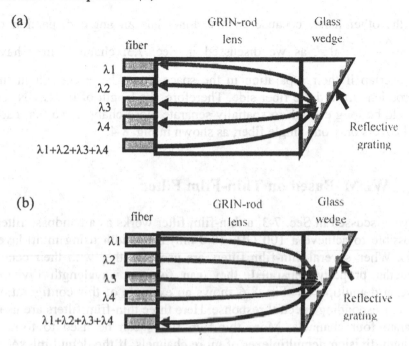

Figure 8-4. Combination of a grating, a glass wedge and a GRIN-rod lens for (a) wavelength division demultiplexing (b) muiltiplexing.

Similar to the optical filter based on grating, channels must not overlap one another when they are multiplexed or demultiplexed. The diffraction beam at a specific direction has the 3 dB linewidth given by λ/N. To prevent signals of neighboring channels from overlapping, the 3 dB linewidth must be much less than the channel spacing. The situation is shown in Fig. 8-5. For a channel spacing of 0.8 nm, we require $\lambda/N <<0.8$ nm, so the number of grating grooves $N >> \lambda/0.8$nm. For $\lambda = 1550$ nm, $N >> 1940$.

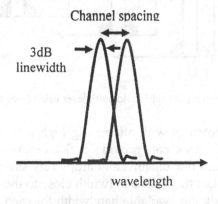

Figure 8-5. Comparison of channel spacing and the 3 dB linewidth of the individual channel.

On the other hand, because each channel has an angle dispersion of

$$\Delta\theta_r = m\frac{\Delta\lambda}{\Lambda}\cos^{-1}\theta_r \ ,$$ as we discussed in Sec. 7.3, channels may have

spatial overlap if their separation in the space is not large enough in the GRIN rod lens toward the fiber side. Therefore, the length of the GRIN rod lens has to be long enough to spatially separate those channels so that each channel covers only one single fiber, as shown in Fig. 8-4.

2.2   WDM Based on Thin-Film Filter

As we discussed in Sec. 7-3, a thin-film filter works as a bandpass filter. It is possible to achieve a 100 GHz (0.8 nm) bandwidth using multi-layer coatings. When several thin-film filters are used together with their center wavelengths properly separated, they can form a wavelength division multiplexer/demultiplexer. Fig. 8-6 shows an example of this configuration and the corresponding spectral response. Here three thin-film filters are used to separate four channels. More thin-film filters can be used to form a wavelength division demultiplexer of more channels. If the demultiplexer is used for the reverse transmission of light, it becomes a multiplexer.

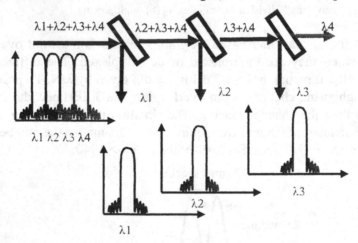

Figure 8-6. A wavelength division demultiplexer using three thin-film filters.

The multi-layer coatings with alternating high refractive index and low refractive index dielectrics can provide a flat region of high reflectivity within the bandwidth of one channel and drops very sharply near the edge of the band. [2,3] This makes the 3 dB bandwidth close to the channel spacing, so significantly increasing the available bandwidth for each channel. Therefore, this type of WDM is superior to the grating-type WDM from this point of view. However, each channel requires a thin-film filter so many thin-film filters are required if the number of channels is large. This will increase the complexity and cost for assembly. In contrast, one grating may provide multiplexing/demultiplexing for many channels, so the grating-type WDM is likely to be more cost-effective for a large number of channels.

Because the thin-film filters in the WDM are used for light traveling in the free space, they have to be properly packaged with fibers for convenient use in optical fiber communication systems. Fig. 8-7 shows an example of the package.

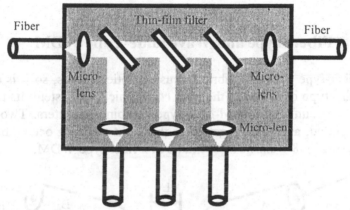

Figure 8-7. A WDM based on thin-film filters is packaged with micro-lenses.

2.3 WDM Based on Fiber Bragg Grating

When a fiber has a periodic variation of refractive index, it gives strong reflection at a wavelength satisfying the condition, $2\Lambda n_{eff} = \lambda_B$, where Λ is the period of the index variation. The reflected wavelength λ_B is the Bragg wavelength. Its function is shown in Fig. 7-27. The reflected beam goes back to the input fiber. The light of reflected wavelength λ_B needs to be spatially separated from the input beam. Therefore, a circulator or a 3 dB coupler is required. The configuration using fiber Bragg gratings and circulators is shown in Fig. 8-8. Two wavelengths are selected out in the configuration shown in this figure. More wavelengths can be selected by cascading more fiber Bragg gratings and circulators. This configuration provides wavelength demultiplexing.

The circulator is a nonreciprocal component. That is, light does not trace its original route when it is sent in from the output end. Thus the configuration shown in Fig. 8-8 does not work as a multiplexer. It has to be reconfigured. The order of the ports in the circulator must be reversed, so the direction of light circulation becomes counter-clockwise, which is opposite to the direction shown in Fig. 8-8.

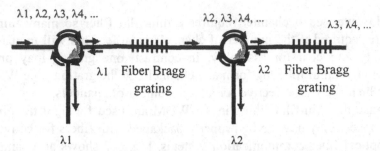

Figure 8-8. A WDM based on fiber Bragg grating and circulator.

2.4   Fiber-Type and Waveguide-Type WDM

A fiber-type WDM is fabricated using optical fibers, so it is convenient to use this type of WDM in the fiber communication system. Its principle of operation is similar to fused-fiber type combiners/splitters. Two fibers are heated, pulled, and fused together so that light coupling occurs between the two fibers. Fig. 8-9 shows a schematic of a fiber-type WDM.

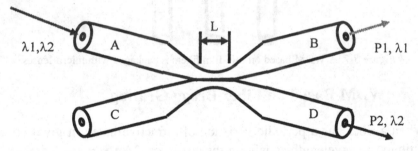

Figure 8-9. A waveguide type WDM that separates two wavelengths.

In contrast to the usual 3 dB fused-fiber combiners/splitters, the coupling coefficients of the fiber type WDM are highly wavelength dependent. The length of the coupling region is designed to have maximum transmission from A to B for $\lambda 1$ and from A to D for $\lambda 2$, as shown in Fig. 8-9. This is made possible by the proper design of the two fibers. According to Eqs. (6-67a) and (6-67b), the coupling coefficients are functions of the wavelength through several factors. The wave vectors k_o, β_1 and β_2 are most important because they have strong dependence on the wavelength. It causes the coupling ratio to vary with the wavelength at a fixed coupling distance. For two identical fibers, the power variation in the two fibers is given by Eqs. (6-70a) and (6-70b). When the coupling coefficients are highly dependent on the wavelength, the power P_1 and P_2 in Eqs. (6-70a) and (6-70b) are a function of wavelength. Therefore, we have

$$P_1(\lambda) = P_{in}(0)\cos^2[\gamma(\lambda)L] \qquad (8\text{-}4a)$$

$$P_2(\lambda) = P_{in}(0)\sin^2[\gamma(\lambda)L] \qquad (8\text{-}4b)$$

The power variation as a function of wavelength is shown in Fig. 8-10. We can design the coupling length so that that $\gamma(\lambda 1)L=2\pi n+\pi/2$ for wavelength $\lambda 1$ and $k(\lambda 2)L=2\pi m+\pi/2$ for wavelength $\lambda 2$, where n and m are integers. Then we have $P_1(\lambda 1) = P_{in}(0)$ and $P_2(\lambda 1) = 0$. In addition, $P_2(\lambda 2) = P_{in}(0)$ and $P_1(\lambda 2) = 0$ for wavelength $\lambda 2$. [4] Therefore it provides wavelength separation.

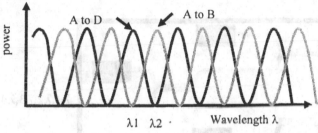

Figure 8-10. Variation of the power of the output1 and the output2 with the wavelength.

The above discussion explains the function of this device for use as a demultiplexer. When the light is delivered in the reverse direction, it will function as a multiplexer. Refer to the WDM shown in Fig.8-9 as an example. If port B is used as the input port and the light with wavelength $\lambda 1$ is input into port B, the power at port A for this wavelength is given by Eq.(8-4a). For the same coupling coefficient and coupling length, port A certainly has the same power as port B if the insertion loss is neglected. On the other hand, if the light with wavelength $\lambda 2$ is input into port D, the power at port A for this wavelength is given by Eq.(8-4b) because they are not in the same fiber. Again, with the same coupling coefficient and coupling length, port A also has the same power as port D for the insertion loss neglected. When wavelength $\lambda 1$ and wavelength $\lambda 2$ are input into port B and port D respectively at the same time, they will both appear at port A. Thus the function of multiplexing is achieved.

The coupling between two fibers can be also achieved by polishing the fiber to remove the cladding material. When the two-polished fibers are put together, their cores in the coupling region are very close, so coupling occurs. For wavelength spacing of 35 nm to 200 nm, channel separation approaching 10 dB to 50 dB may be achieved in terms of power ratio between channels.[5]

Multiplexers with more than two ports can be constructed by cascading several stages of the basic 2 x 1 multiplexers. Fig. 8-11 shows a 4 x 1 multiplexer formed by three 2 x 1 multiplexers.

The design concept of a waveguide-type WDM is the same as the fiber-type waveguide. The waveguides are physically brought very close together in the coupling region, so coupling takes place, like the situation in the waveguide combiners/splitters. However, the coupling region for WDM is designed such that the coupling coefficients depend strongly on the wavelength. Therefore, the discussion for the fiber-type WDM can be applied here and both multiplexing and demultiplexing functionality can be achieved.

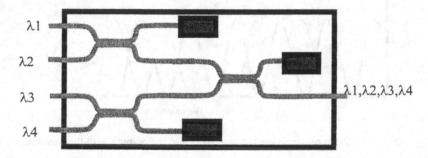

Figure 8-11. A 4 x 1 multiplexer formed by three 2 x 1 multiplexers.

Packages for light delivery from fibers to the waveguides are also necessary for use in the optical fiber communication system. Fig. 8-12 schematically shows an example of the package.

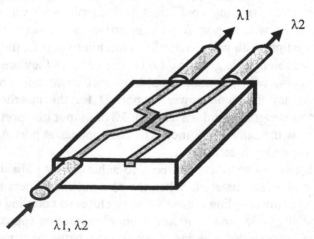

Figure 8-12. Package for light delivery between fibers and the waveguides in the waveguide-type WDM.

Multiplexers with a large number of ports can also be constructed by cascading several stages of the basic 2 x 1 multiplexers. In addition, 2 x 1 waveguide-type multiplexers can be monolithically integrated on the same substrate, making the formation of multi-port WDM very easy. Furthermore, techniques of semiconductor processing are applied in the fabrication, so mass-production is possible at a very effective cost. For a very large number of ports, the waveguide-type WDM is advantageous over fiber-type WDM from the viewpoint of fabrication and cost.

2.5   WDM Based on Mach-Zehnder Interferometer

The Mach-Zehnder interferometer can be used as a WDM. Its configuration is shown in Fig. 8-13. The waves at two wavelengths are input into the same port. The input light is then split equally to path1 and path2 by the 3 dB splitter at the input side. Path2 is longer than path1 for a length of ΔL. The length ΔL is intentionally designed to have the following conditions.

$$n_{eff} \Delta L = n\lambda 1 \tag{8-5a}$$

$$n_{eff} \Delta L = (m+1/2)\lambda 2 \tag{8-5b}$$

where n_{eff} is the refractive index that light experiences in both path1 and path2; n and m are integers.

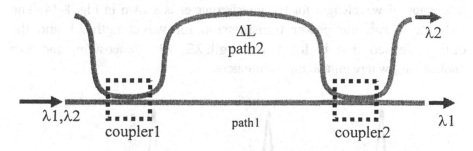

Figure 8-13. WDM based on Mach-Zehnder interferometer.

After passing through path1 and path2, the waves combine again at the second 3 dB combiner/splitter, also known as a 3 dB coupler. For wavelength $\lambda 1$, both waves from the two paths are in phase according to Eq.(8-5a) right before they enter the 3 dB coupler. For wavelength $\lambda 2$, both waves from the two paths are out of phase according to Eq.(8-5b). From our discussion in Sec. 7.2.2.3 - Sec. 7.2.2.4, we know that the wave at $\lambda 1$

experiences constructive interference at port 1 and wave at λ2 experiences constructive interference at port 2. Therefore, they are separated.

When light waves with wavelengths λ1 and λ2 are sent in separately from each input port, they combine at the first 3 dB coupler. At the output of the first 3 dB coupler, the phase issues are similar to the discussion on bulk-type combiners/splitters in Sec. 7.2.2.3. The phase difference between the two outputs of the 3 dB coupler is not the same for the two wavelengths. After they travel through path1 and path2, the path difference causes the waves at both wavelengths to be in phase right before they enter the 3 dB coupler. They then both exit the same output port, thereby functioning as a multiplexer.

2.6   WDM Based on Fabry-Perot Interferometer

In Sec. 7.3.2.1, we discussed the use of a Fabry-Perot interferometer as a tunable filter. The Fabry-Perot interferometer is characterized by the transfer

function $T_{FPF} = \dfrac{(1-R)^2}{(1-R)^2 + 4R\sin^2[(\omega - \omega_0)d/n]}$. T_{FPF} represents the ratio

of power transmission through the interferometer. The transfer function and the peaks periodically appear in the wavelength domain with the period

$\Delta\lambda = \dfrac{\lambda^2}{2nd\cos\theta}$, which is also the free spectral range (FSR). θ is the angle of

light incidence on the Fabry-Perot interferometer. The power transmission as a function of wavelength for the interferometer is shown in Fig. 8-14. The red line shows the power transmission for wavelength λ1 and the orange dashed line is for wavelength λ2. $\lambda1 = 2nd\cos\theta/m_1$ and $\lambda1 = 2nd\cos\theta/m_2$, where m_1 and m_2 are integers.

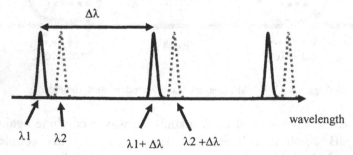

Figure 8-14. The power transmission as a function of wavelength for the interferometer. The red line is for wavelength λ1 and the orange dash line is for wavelength λ2.

Fig. 8-14 shows that the light at wavelength λ1 transmits through the interferometer, while other wavelengths between λ1 and λ1 + Δλ are reflected. Therefore, if the light at several wavelengths, which include λ1 and other wavelengths within the FSR between λ1 and λ1 + Δλ, is incident on the Fabry-Perot interferometer, the wavelength λ1 will separate from other wavelengths. Then if we cascade several Fabry-Perot interferometers with their peak-transmission wavelengths properly separated, we are able to demultiplex those channels of different wavelengths, as shown in Fig. 8-15. It is obvious that it will work as a multiplexer when the light is delivered in directions opposite to those shown in Fig. 8-15.

Similar to thin-film filters, the Fabry-Perot interferometers are used for light traveling in the free space, so they have to be properly packaged with fibers. It is thus necessary to use lenses to couple light into and out of fibers and to have a collimated beam propagating in the package.

For the channel spacing δλ, the number of channels we can put within the FSR is Δλ/ δλ. On the other hand, the 3 dB bandwidth for each transmission line $\dfrac{c}{2nd}\dfrac{1-R}{R^{1/2}}$ should be much narrower than the channel spacing δλ in order to reduce crosstalk between neighboring channels.

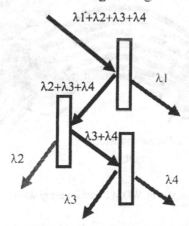

Figure 8-15. A WDM based on Fabry-Perot interferometer.

2.7 WDM Based on Array Waveguide Grating (AWG)

Array waveguide gratings (AWG) [6-8] are formed with an array of waveguides. Array waveguide gratings are also named Phased Arrays (PHASARs) and Waveguide Grating Routers (WGRs). [6, 9] They are planar devices. A schematic of the WDM based on array waveguide gratings is shown in Fig. 8-16. It has three major regions:
1. the free expansion region behind the input waveguide, being an N x M star coupler;
2. array waveguide gratings;

3. M x N star coupler.

The principle of operation for the AWG is similar to that of the diffraction grating. It utilizes the interference effect of the light wave. The interference leads to the light at wavelength λ1 having constructive interference at the output waveguide of port 1, but destructive interference at other output waveguides. Similarly, the light at wavelength λ2 has constructive interference at the output waveguide of port 2, but destructive interference at other output waveguides. Other wavelengths behave in the same way. The detail of the principle is explained in the following.

The first star coupler is designed such that the right edge of this region facing the arrayed waveguides coincides with the wavefront of the wave freely expanding from the aperture of the input waveguide. The wave exiting the input waveguide expands freely and arrives at the arrayed waveguides. The wave then enters those waveguides with optical power from the input waveguide equally split to the M waveguides.

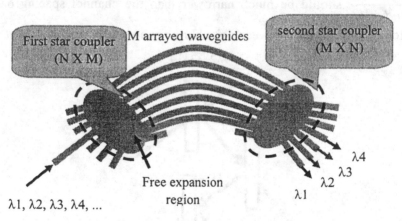

Figure 8-16. A schematic of WDM based on array waveguide gratings.

The arrayed waveguides are intentionally designed with lengths varying in a decreasing order and adjacent waveguides differing in length by ΔL. The waves traveling in those waveguides then have a phase difference $\Delta\phi_w$ between the adjacent waveguides when they leave the waveguides.

$$\Delta\phi_w = 2\pi\frac{\Delta L}{\lambda} \tag{8-6}$$

After leaving the waveguides, they propagate further to the output waveguides. The distances from the exits of the waveguides to the individual output port may be different. Thus there is also an additional path difference Δl between the waves in adjacent paths. The situation is illustrated in Fig.8-

17. The path for the light propagating from waveguide 1 to port 1, from waveguide 2 to port 1, and from waveguide M to port 1 are shown. The lengths of those paths are obviously different. The actual path difference depends on the geometrical structure of the star coupler and the relative positions among the arrayed waveguides and the output waveguides. In fact, this situation is similar to the diffraction grating with the incidence angle equal to zero. For the situation where the arrayed waveguide is far apart from the output waveguide, we can apply the path difference derived for Eq. (6-60) in Sec. 6.4.1. Therefore,

$$\Delta l = \Lambda \sin\theta_r \tag{8-7}$$

where θ_r is the diffraction angle viewing from each output waveguide.

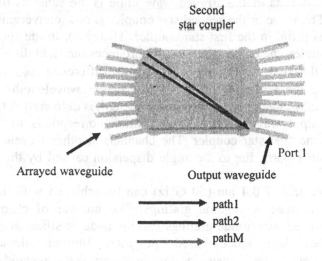

Figure 8-17. Travelling paths of waves in the second star coupler in the second star coupler of AWG-based WDM.

The path difference Δl also causes a phase difference $\Delta\phi_c$ to the waves between the adjacent paths.

$$\Delta\phi_c = 2\pi\frac{\Delta l}{\lambda} \tag{8-8}$$

Therefore, the total phase difference for light between the adjacent paths is $\Delta\phi_w + \Delta\phi_c$. To have constructive interference at the port 1 of the output waveguides, the condition is

$$\Delta\phi_w + \Delta\phi_c = 2\pi(\frac{\Delta L}{\lambda 1} + \frac{\Delta l}{\lambda 1}) = 2m\pi \qquad (8\text{-}9)$$

where m is an integer. The wavelength λp has constructive interference at the output port p.

$$\Delta\phi_w + \Delta\phi_c = 2\pi(\frac{\Delta L}{\lambda p} + \frac{\Delta l}{\lambda p}) = 2m\pi \qquad (8\text{-}10)$$

In practice, the second star coupler is usually designed to have the mirror-imaged shape of the first star coupler. When the phase difference $\Delta\phi_w$ caused by the arrayed waveguides is in multiples of 2π, the shape of the wavefront at the exits of the arrayed waveguide is the same as that at the entry points. The wave in the second star coupler is now converging instead of diverging as it did in the first star coupler. Therefore, in the space of the second star coupler, the converging wave reproduces the field distribution in the space of the first star coupler. The phase difference $\Delta\phi_w$ is usually designed to be the multiples of 2π for the center wavelengths of those channels. Thus the channel at the center wavelength is delivered to the center port of the output waveguides whose position corresponds to the input waveguide in the first star coupler. The channels at other wavelengths are delivered to other ports due to the angle dispersion caused by the different wavelengths. [6]

Channel spacing of 0.4 nm (50 GHz) can be achieved with the WDM based on the arrayed waveguide gratings. The number of channels can exceed 48. Arrayed waveguide gratings can be made of silica-on-silicon or other materials like silicon-based polymer, lithium niobate, and semiconductor, etc. [9] The propagation loss in the arrayed waveguides can be less than 0.1 dB/cm. Light coupling between the fiber and the input/output waveguides of AWG-based WDM can be below 0.5 dB. The most severe loss is the junction loss between the waveguides and the free propagation region of the first and the second star couplers, 1-1.5 dB for each junction and 2-3 dB for the total device.

3.   OPTICAL ADD/DROP MULTIPLEXER/ DEMULTIPLEXER (OADM)

In a WDM optical communication system, not every channel has the same destination. It is very common that one channel has arrived at a final node, while other channels are destined to other places. It is like many passengers in a train running from New York to San Francisco. In the middle of the journey, one passenger arrives at his destination in a small town and gets off the train. Others want to continue their long trip to the west coast. Conventionally, the train has to stop at this town, letting this passenger get off. All other passengers have to stop at this town, too, although they may not want to. These stops certainly delay the arrival time of the train at the west coast. It would be nice if the train did not have to stop and the passengers could still safely get off the train at their intermediate destinations. This is actually possible for data transmission in the WDM optical communication system using optical add/drop multiplexers/ demultiplexers (OADM).

The functional block of an OADM is shown in Fig. 8-18. Several signal channels are delivered in the optical system. When they arrive at the OADM, only the signals at wavelength λ3, indicted by the red color, are taken out, while other signals continue their transmission without stop. This wavelength can then transmit to a different direction from those at wavelengths λ1, λ2, λ4. On the other hand, this OADM can also add signals at wavelength λ3, which were originally not grouped together with the other wavelengths. The added signals at wavelength λ3 are indicated by the blue color in Fig. 8-18. The added signals at wavelength λ3 are generally not the same as the dropped signals at wavelength λ3. Returning to our train analogy, this is like several passengers getting off the train at an intermediate destination with several new passengers taking their place without requiring that the train make a stop.

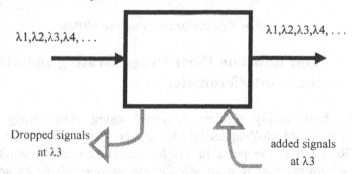

λ1,λ2,λ3,λ4, . . .   λ1,λ2,λ3,λ4, . . .

Dropped signals
at λ3   added signals
at λ3

Figure 8-18. Function block of an optical add-drop multiplexer/demultiplexer.

There are many ways to accomplish the function depicted in Fig.8-18. The simplest way is the use of two WDMs configured as shown in Fig. 8-19. The WDM DEMUX separates those channels of different wavelengths. The channel of wavelength λn is then taken out of the system, while other channels are directly connected to the inputs of WDM MUX. Therefore signals at other wavelengths transmit though without interruption. On the other hand, if the WDM MUX has another unused input channel, we can add signals at wavelength λn to this channel, multiplexed by the WDM MUX. The added signals then transmit together with the other channels in the WDM system.

In the following sections, some specific examples of OADMs will be given. There are more possible configurations of OADMs than the given examples. Since many new configurations can be designed using the principles of optics and the concept illustrated in Fig. 8-19, they will not be elaborated on here.

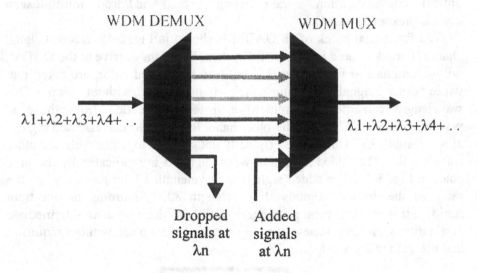

Figure 8-19. An OADM using two WDMs.

3.1   OADM Based on Fiber Bragg Grating and Mach-Zehnder Interferometer

OADM functionality can be achieved using fiber Bragg gratings configured with a Mach-Zehnder Interferometer. The configuration is shown in Fig.8-20. Each of the paths in the Mach-Zehnder interferometer has a fiber Bragg grating that is designed for the wavelength to be added and dropped.

In Fig.8-20, all wavelengths are sent to the same input port of the first 3 dB coupler. Their power are split equally to path1 and path2, but their phases are different, as we discussed in Sec. 7.2.2.3 and Sec. 7.2.3. The wave at wavelength λn is reflected by the fiber Bragg grating for both path1 and path2. The reflected wave travels back to the first 3 dB coupler. When they combine in the first 3 dB coupler, the phase difference between the waves in path1 and path2 causes the reflected wave at wavelength λn to go to the input port 2. Thus the function of signals dropping for the wavelength λn is achieved.

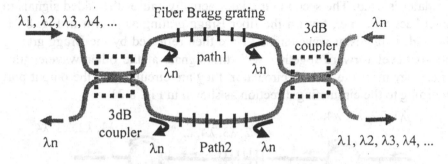

Figure 8-20. OADM based fiber Bragg gratings and Mach-Zehnder Interferometer.

The other wavelengths continue to travel through the two paths and then combine at the second 3 dB coupler. With proper adjustment of the phase difference due to the path difference between path1 and path2, they exit at output port 2.

The signals at wavelength λn can be added at output port 1, as shown in Fig. 8-20. This added wave at wavelength λn is split into the two paths with equal power and different phases, as we discussed in Sec. 7.2.2.3 and Sec. 7.2.3. The waves in the two paths are then reflected by the fiber Bragg grating and travel back to the second 3 dB coupler. When they combine in the second 3 dB coupler, the phase difference between the waves in path1 and path2 causes the reflected wave at wavelength λn to go to a different port from the one it is entered. Thus the signals at λn are added to the same output port for other wavelengths.

3.2 OADM Based on Fiber Bragg Grating and Circulator

A simpler configuration than the previous one can be obtained using one fiber Bragg grating and two circulators. It is shown in Fig. 8-21. The fiber Bragg grating is designed for the wavelength to be added and dropped. That is, it gives strong reflection at the wavelength λ1 that satisfies $2\Lambda n_{eff} = \lambda1$, where Λ is the period of the index variation in the fiber. The reflected wavelength λ1 is the Bragg wavelength. Light at several wavelengths is delivered into the fiber. The Bragg grating in the fiber causes the light at wavelength λ1 to be reflected. The reflected beam goes back to the input fiber. It then needs to be spatially separated from the input light. Therefore, a circulator is required.

Signals at other wavelengths continue to travel in the fiber as shown in Fig. 8-21. If the signals at wavelength λ1 are to be added, the second circulator is used. The second circulator actually causes the added signals to travel backward, i.e., toward the fiber Bragg grating as shown in Fig. 8-21. The added signals at wavelength λ1 are then reflected by the Bragg grating and so travel forward together with the signals at the other wavelengths. When they meet the second circulator, they are circulated to the output port according to the circulating direction as shown in Fig. 8-21.

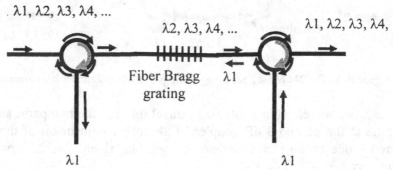

Figure 8-21. OADM based fiber Bragg gratings and circulator.

3.3   OADM Based on Fabry-Perot Interferometer

The discussion on the use of Fabry-Perot interferometers for WDM given in Sec. 8.2.6 greatly helps us understand the OADM based on the Fabry-Perot interferometer. When the light at several wavelengths is incident on the Fabry-Perot interferometer designed for a certain wavelength, say λ1, this wavelength passes through and other wavelengths are reflected. Thus the wavelength λ1 separates from other wavelengths. This is the function desired for OADM. Therefore, the OADM based on the Fabry-Perot interferometer can be easily configured, as shown in Fig. 8-22.

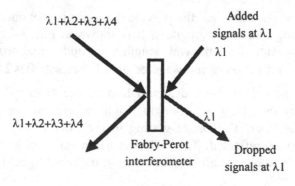

Figure 8-22. OADM based on Fabry-Perot interferometer.

3.4   Tunable OADM

The add-drop channel of the OADM can be made tunable. For example, the selected channel λ1 of the OADM based on Fabry-Perot interferometer discussed previously becomes tunable if the Fabry-Perot interferometer is actually a Fabry-Perot tunable filter discussed in Sec. 7.3.2.1. A general schematic using tunable filters to make the add-drop channel of the OADM tunable is shown in Fig. 8-23. With an understanding of tunable filter design, it is straightforward to realize its operating principle.

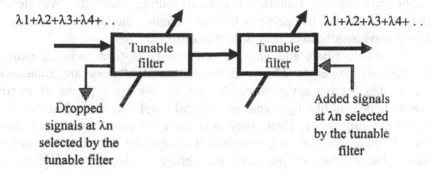

Figure 8-23. Tunable OADM using tunable filters.

The tunable OADM can also be made by using optical switches or wavelength routers. They will be described after the following discussion of optical switches and wavelength routers. The tunability of the add-drop channel of the OADM can be easily understood with the knowledge of these two components.

4.   OPTICAL SWITCH, OPTICAL CORSSCONNET AND WAVELENGTH ROUTER

Switches are required in optical communications systems just as they are in electronic telecommunications systems,. They can simply serve to turn on or off a signal in a communication path, switch a signal from one path to another, or even perform more complex functions. In conventional systems, each signal channel transmits on its own physical wire. When signals have to be directed to different destinations, we only need to connect each wire to its own destination. However, in WDM communication systems, many channels of signals are simultaneously transmitting in a single fiber. Physical separation of those channels by connecting one wire for each channel to its own final destination is not possible. Then the issues for directing those channels to different destinations are not trivial. In some cases, switches serving simple functions are sufficient. In other cases, connections among complicated optical routes are required. The switching requirements and complexity in optical communication systems can be much more complicated than the electronic telecommunication systems.

In general, optical crossconnects, routers, etc. are similar to optical switches. They all serve similar functions although sometimes their distinctions are not clear. When the number of switching ports used in a WDM system is large, we usually call such optical switches optical crossconnects, optical routers, or optical routing elements. Wavelength routers are complex components that reconfigure the optical channels of different wavelengths to be transmitted in different fibers.

Like tunable filters and variable attenuators, optical switches usually need external power to carry out their functions. They are sometimes categorized as active components. However, they are different from true active components like light sources, optical amplifiers and photodetectors in the following ways. First, they involve no electro-optical and opto-electrical conversion. Second, they do not change the formats of the optical signals. That is, they do not have the ability to change the information content bit by bit. For the most part, they apply the principles of passive optical components instead of the active components discussed in earlier chapters. Therefore, we discuss them in this chapter.

The functions of switching can be achieved using electronic switches, as shown in Fig. 8-24(a). There the optical signals are converted to electronic signals and then switched using electronics. Afterwards, the electronic signals are converted back to optical signals and transmitted in optical fibers. These conventional ways involving the O-E and E-O conversion are seldom used now. All optical switches have been popular for their mature technology and better performance. Their functional block is plotted in Fig.8-24 (b) for comparison with switching using electronics. The discussion in this part is for all optical switches.

(a)

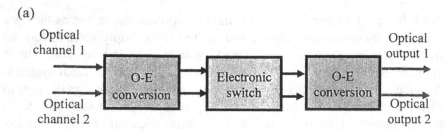

(b)

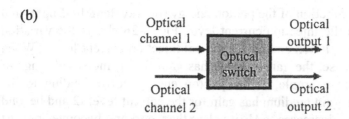

Figure 8-24. (a) Optical switch accomplished by conventional electronic switch. (b) Optical switch accomplished by optical ways.

4.1 Optical switch

Switches can be classified according to their port numbers. There are
- on/off switch;
- 1 x 2 switch;
- 1 x N switch;
- 2 x 2 switch;
- N x N switch.

To make the explanation easier, we will classify optical switches/crossconnects as follows. Those with a single input port are called optical switches and will be explained in this subsection. Those with multiple input ports, including the 2 x 2 switch and N x N switches are called optical crossconnects and will be explained in next subsection.

4.1.1 On/off Switch

The function of an on/off switch is very simple. When it is on, the light signals are allowed to pass through the path. When it is off, no light signals pass through. Its function is shown in Fig.8-25. It is often used in test equipment for the complete isolation between the sources and the receivers.

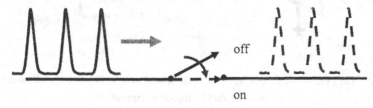

Figure 8-25. The function of an on-off switch. (The optical signals pass through when the switch is on.)

A semiconductor optical amplifier (SOA) is a good candidate for use as an on/off switch. The gain medium of a SOA was explained in Chapters 2 and 4. Its gain is a function of the photon energy (or wavelength of light) and depends highly on the injection current level. Fig. 8-26 shows the variation of the gain with the photon energy for several injection current levels. When current levels increase, the gain medium has gain for an increased range of photon energy. For example, at the photon energy corresponding to the wavelength λ1, the gain medium has gain for the current level I2 and beyond. As the current level decreases to I1, it is less than zero and becomes loss. At the photon energy corresponding to the wavelength λ2, the gain medium still has loss for the current level I2. When the SOA has gain, it will allow the light signals to pass through. When the SOA has loss, it absorbed the power of the light signals and leave out no optical signals. Therefore, the SOA works as an on/off switch and is controlled by the current or the voltage. Its functional block is represented by the symbol in Fig. 8-27.

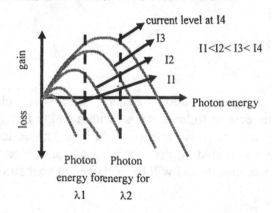

Figure 8-26. Gain as a function of photon energy for different current levels.

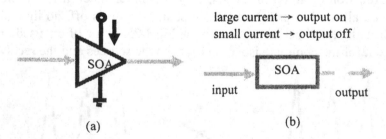

Figure 8-27. (a) Symbol of a SOA working as an optical switch. (b) The on/off switch of SOA is controlled by injection current.

4.1.2   1 x 2 Switch

The 1 x 2 switch connects the path from the input to either output 1 or output 2, but not both. Its function is shown in Fig.8-28. Several schemes can be used to form 1 x 2 switches. A few examples are explained in the following. Fig. 8-29 shows a 1 x 2 switch using the waveguide coupler fabricated on material with electro-optic effect like LiNbO$_3$. The electro-optic effect causes the refractive index of the waveguide to vary with the applied voltage, as explained in Chapter 3. On the other hand, the coupling coefficients depend on the refractive index, so the coupling coefficients vary with the applied voltage. According to Eqs.(6-70a) and (6-70b), the power at the two outputs depends on the coupling coefficients, so they are controlled by the applied voltage. With proper design, the coupling goes to either output port 1 or port 2, depending on the applied voltage.

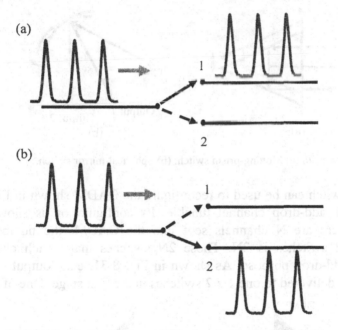

Figure 8-28. 1 x 2 switch: (a) connection between input and output 1; (b) connection between input and output 2.

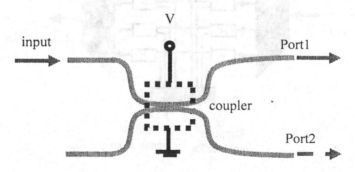

Figure 8-29. 1 x 2 switch based on a waveguide coupler.

Switching can be achieved using mechanical movement. Fig.8-30(a) shows a moving-prism switch, the lateral movement of the prism causes the reflected beam to be directed toward either output 1 or output 2. Fig.8-30(b) shows a spherical-mirror switch, the rotation of the mirror causes the reflected beam to be directed toward either output 1 or output 2. The pivot of rotation coincides with the incident point of the light path.

Many other types of 1 x 2 switches can be constructed using various schemes. With the knowledge of optical principles discussed previously, their functions can be easily predicted. The 1 x 2 switch can be used in the reverse direction. That is, we can use it as a device with 2 inputs and 1 output.

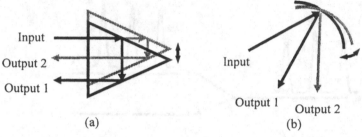

Figure 8-30. (a) Moving-prism switch. (b) Spherical-mirror switch.

The 1 x 2 switch can be used to reconfigure the OADM shown in Fig. 8-19, making the add-drop channel tunable. Its configuration is shown in Fig.8-31. If there are N channels sent into the OADM, the number of required 1 x 2 switches is 2N. Those 2N switches make each channel available for add-drop purpose. As shown in Fig. 8-31, each output of the left DEMUX is delivered to one 1 x 2 switches in the first stage. One of the

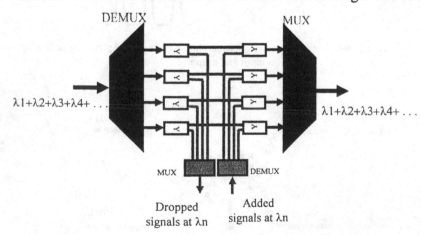

Figure 8-31. A tunable OADM using 1 x 2 switches and WDMs.

outputs in the first-stage 1 x 2 switch is delivered to the second-stage 1 x 2 switch. The other output in the first-stage is delivered to another MUX shown in the bottom of the figure. The second-stage 1 x 2 switches are used in a reverse order. The outputs of the second-stage are sent to the right MUX. One of the inputs comes from the first-stage 1 x 2 switch, while the other input is from the DEMUX shown in the bottom of Fig. 8-31. These 1 x 2 switches then make the OADM tunable. In addition, in this configuration, the number of added or dropped channels is not limited to one.

4.1.3   1 x N Switch

The 1 x N switch is the extension of 1 x 2 switch. It can be formed either by cascading several stages of 1 x 2 switches or directly increasing the number of ports at the output. Take the 1 x 2 switches described above as examples. Cascading two stages of the 1 x 2 switch based on the waveguide coupler, we can form a 1 x 4 switch. Directly changing the number of output ports of the moving prism switch, we are able to make a 1 x 5 switch, as shown in Fig. 8-32. Because the light has to be coupled into optical fibers, the precision of alignment is very critical. For switches using mechanical movement, the bi-state positions of 1 x 2 switch makes the precision much easier to achieve than the 1 x N switch.

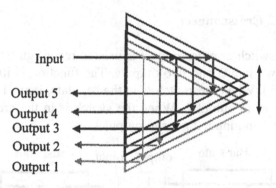

Figure 8-32. 1 x 5 moving-prism switch.

1 x N switches are often used for the test and measurement of optical components for communication.[10] In test and measurement, many devices may have to be characterized. 1 x N switches enable us to route the test paths without manual reconfiguration. This helps avoid the connection mistakes when we have to test many devices. A possible configuration of testing setup using these switches is shown in Fig. 8-33.

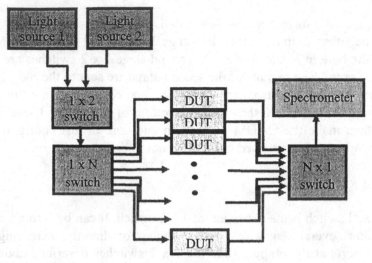

Figure 8-33. A configuration of testing setup using 1 x 2 and 1 x N switches. (DUT: device under test)

4.2   Optical Crossconnect (OXC)

In this subsection, we will discuss the 2 x 2 and N x N switches that have multi-port inputs.

4.2.1   2 x 2 Crossconnect

The 2 x 2 switch or crossconnect is used to reconfigure the optical routes between the two inputs and two outputs. The function is illustrated in Figs. 8-34 (a) and (b). When the switch is in the bar state, input 1 goes to output 1 and input 2 goes to output 2. When the switch is in the cross state, input 1 goes to output 2 and input 2 goes to output 1.

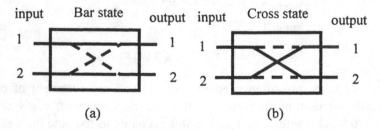

Figure8-34. 2 x 2 switch (crossconnect): (a) bar state; (b) cross state.

The directional couplers and Mach-Zehnder interferometers discussed in Chapter 7 are natural 2 x 2 switches. The 2 x 2 optical switch can also be made from mechanical movement, usually called an optomechanical switch. Fig. 8-35 shows a mirror-based optomechanical switch.

Four SOAs can be used to form a 2 x 2 optical switch, as shown in Fig. 8-36. However, such a configuration has a slightly different function from the other types of 2 x 2 switch. It does not have the bar state or the cross state. Only one input has transmission to either output at one time. When the

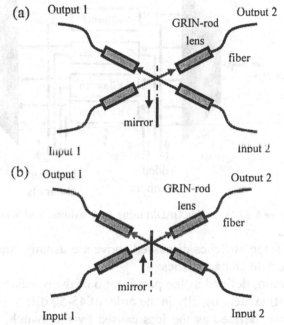

Figure 8-35. A mirror-based optomechanical switch: (a) cross state with mirror away from the light path; (b) bar state with mirror blocking the light path.

SOAs (a)-1 and (b)-1 are on and the other two SOAs are off, only input A1 goes to output B1. Please note that input A2 does not go to output B2. When the SOAs (a)-1 and (b)-2 are on and the other two SOAs are off, input A1 only goes to output B2. Again, input A2 does not go to any outputs. For the transmission of the input A2 to either output, we turn on and off those SOAs accordingly.

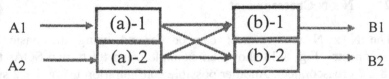

Figure 8-36. A 2 x 2 optical switch using 4 SOAs.

The 2 x 2 switch can also be used to provide the OADM shown in Fig. 8-19 with tunable add-drop channels. Its configuration is shown in Fig.8-37.

Its function can also be traced in a similar way to the tunable OADM shown in Fig.8-31. Again, this configuration provides not only the tunability of add-drop channels, but also multiple add-drop channels.

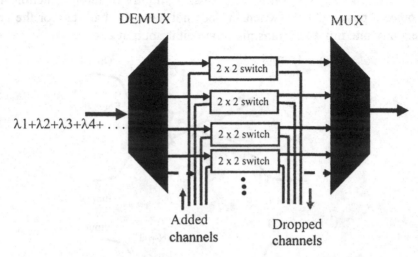

Figure 8-37. A tunable OADM using 1 x 2 switches and WDMs.

The single stage switches discussed above are usually specified with the following important characteristics:

– extinction ratio, defined as the power ratio of the on-state to the off-state for the on-off switch, usually in the order of 45-50 dB;

– insertion loss, defined as the loss caused by the switch, usually on the order of 0.5 dB;

– crosstalk, defined as the power ratio of the output from the desired input to the undesired input, typically on the order of 80 dB;

– switching time, ranging from a few ms to a few ns. For packet-switching applications, the switching time has to be a few ns or even ps. This can be achieved using SOA or electro-optic effects. Switching times for optomechanical switches are often in the range of a ms.

4.2.2 N x N Crossconnect

The N x N crossconnect can be formed using the basic 2 x 2 crossconnects. Fig. 8-38 shows an example using thirty-two SOAs to form one 8 x 8 crossconnect. Another possible configuration using 2 x 2 switches is shown in Fig. 8-39. An N x N crossconnect can also be formed directly using 1 x N switches, as shown by the example of the 4 x 4 switch. Both provide the flexibility of forming an N x N crossconnect with arbitrary value of N.

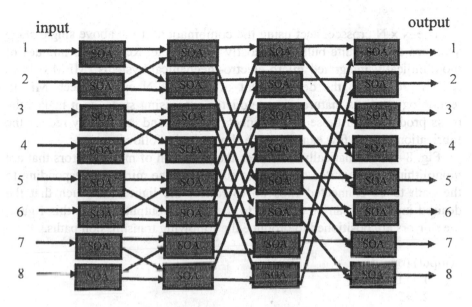

Figure 8-38. 8 x 8 crossconnect formed by 32 SOAs.

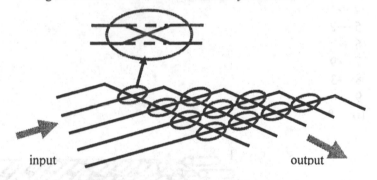

Figure 8-39. 5 x 5 switch using ten 2 x 2 switches.

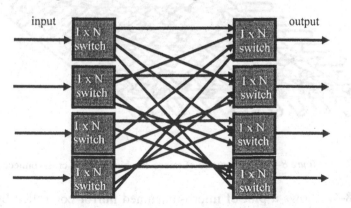

Figure 8-40. 4 x 4 switch using eight 1 x 4 switches.

The N x N crossconnect using the combination of the above single-stage switches can be quite bulky and costly to fabricate. Recently techniques of monolithically integrated micro-electro-mechanical systems (MEMS) have shown promise in the development of the N x N crossconnect. MEMS technology utilizes mature semiconductor processing steps, so it makes the mass production of those crossconnects possible and can greatly reduce the fabrication cost. A few examples are shown in the following.

Fig. 8-41 schematically illustrates the operation of micro-mirrors that are monolithically integrated on a single substrate. The mirror corresponding to the path to be connected flips up to reflect the light beam such that the desired input port can be directed to the desired output port. In this figure, the mirrors are positioned to produce the following transmission paths.

(input) (ouput)
$$1 \rightarrow 1$$
$$2 \rightarrow 5$$
$$3 \rightarrow 4$$
$$4 \rightarrow 10$$
$$5 \rightarrow 2$$
$$6 \rightarrow 7$$
$$7 \rightarrow 9$$
$$8 \rightarrow 3$$
$$9 \rightarrow 6$$
$$10 \rightarrow 8$$

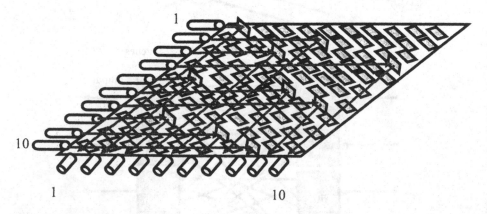

Figure 8-41. A schematic of micro-machined N x N crossconnect.

Fig. 8-42 shows a plot of micro-machined mirror controlled by moving hinges that are also micro-machined.[11, 12] Another example of the flipping mirrors is shown in Figs. 8-43 (a) and (b). The flipping action of the mirror is controlled by the current flowing in the individual pixel and the magnet. Fig.8-43 (a) shows the monolithically integrated mirrors to serve for 10 x 10 crossconnect and Fig.8-43 (b) shows the movement of a single mirror. [13] In the above configuration, we need N x N flipping mirrors, which can greatly reduce the fabrication yield when the port number is large. Another method which requires significantly fewer mirrors is developed to serve for the crossconnect with the same port number. [12] Each micro-mirror is allowed to rotate in two dimensions. In this configuration, either N or 2N micro-mirrors are required for an N x N crossconnect. When the input and the output are switched in the same direction, each port-to-port connection goes through three reflections, two from micro-mirrors and one from a fixed-angle mirror. This requires 2N micro-mirrors. If the input and the output are switched in different directions, only N micro-mirrors are required. Fig. 8-44 (a) illustrates light directions reflected by a 3 x 3 crossconnect using three mirrors. The control of mirror using micro-machined actuators is shown in Fig. 8-44 (b). Although this configuration requires significantly fewer micro-mirrors, it requires analog voltages to control each micro-mirror in order to precisely rotate the mirror and thereby guide the light beam into the fiber. The precise control of those mirrors is a difficult issue.

There are other methods to fabricate N x N crossconnects like bubble switching, [14] liquid crystals, [15, 16] silica-on-silicon, [17] polymer, [18] and so on.

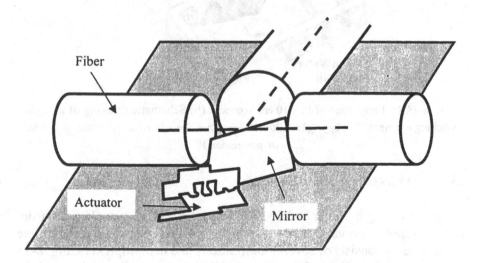

Figure 8-42. A plot of micro-machined mirror used for crossconnect. (mirror and actuator for controlling mirror are shown).

(a)

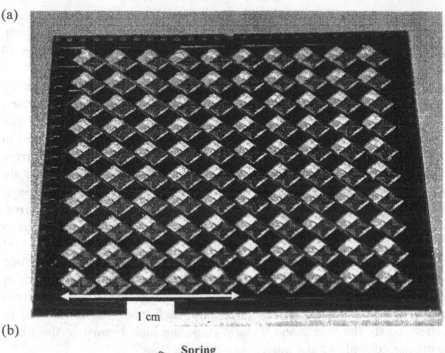

1 cm

(b)

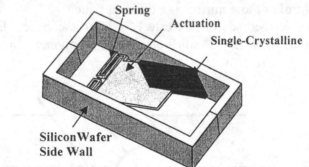

Figure 8-43. (a) Photograph of 10 x 10 crossconnect. (b) Schematic drawing of a single switching element. [13] *(Copyright © 2003. Umachines, Inc. All rights reserved. Reprinted with permission.)*

4.3   Wavelength Routers

The wavelength router receives signals of different wavelengths from multiple inputs, reorganizes those signals and delivers them to multiple output ports. It consists of several multiplexers and demultiplexers. Fig. 8-45 shows its configuration. From Fig. 8-45, we can easily understand its function. It first demultiplexes the different wavelengths from each input. Those different wavelengths are routed with different combinations. Each new combination of wavelengths is multiplexed again and delivered to an output port. Fig. 8-45 shows a 4 x 4 non-configurable wavelength router. It is non-configurable because the combination of the wavelengths delivered to each output is fixed. The routing of those wavelengths is determined by the hardware connection between the demultiplexer and the multiplexer. [19]

(a)

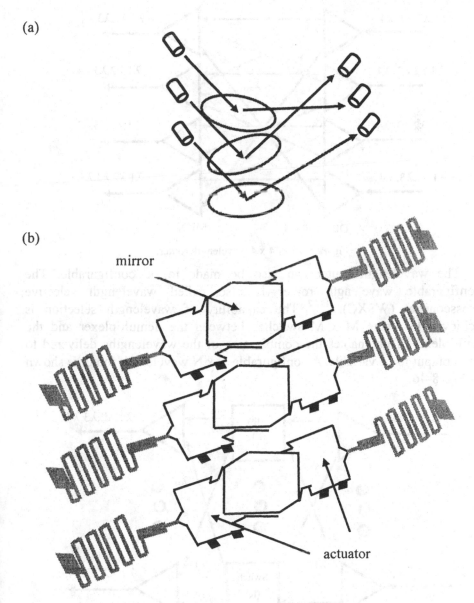

(b)

mirror

actuator

Figure 8-44. (a) A schematic of 3 x 3 crossconnect using three mirrors. (b) Control of micro-machined mirrors using micro-machined actuators.

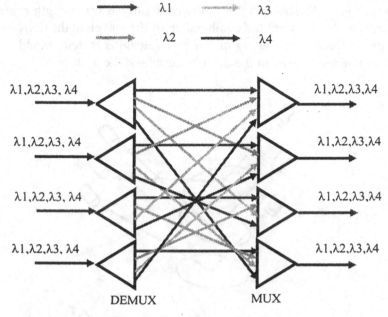

Figure 8-45. A 4 x 4 wavelength router.

The wavelength router can also be made to be configurable. The configurable wavelength router is also called wavelength selective crossconnect (WSXC). [19-22] The capability of wavelength selection is achieved by using M x M switches between the demultiplexer and the multiplexer. This enables the combination of the wavelengths delivered to each output to be variable. A configurable N x N wavelength router is shown in Fig. 8-46.

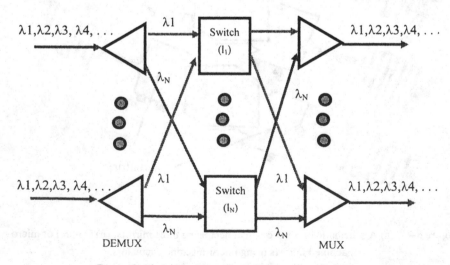

Figure 8-46. An N x N configurable wavelength router.

5. WAVELENGTH CONVERTOR

The optical switches or crossconnects discussed above switch information from one channel or one place to another through hardware wiring or routing. It is also possible to change information from one channel to another without hardware switches and routers. The technique that achieves this is "wavelength conversion", where the information is exchanged from one wavelength to another without reconfiguring the fibers. The exchange of information with wavelength conversion can occur among channels that are carried by the same fiber. Wavelength conversion is also necessary for the following situations.

First, the first generation of communication systems transmit signals at a wavelength of 1.31 μm. Most newer communications networks operate at 1.55 μm. When we want to exchange signals in current systems with first-generation systems, we have to convert our signals from 1.55 μm to 1.3 μm and vice versa.

Second, the communication systems may be provided by different companies using different wavelengths. To make the signals transparent among users that subscribe different systems, wavelength conversion provides the capability to exchange signals among different systems.

Third, two local systems serving two distant towns are initially established at the same wavelength. At a later time, when both are connected to the same WDM system serving a large area, their signals will conflict with one another in the new WDM system. To avoid the conflict, one of the local systems must convert its signals to a new wavelength before entering the WDM system. This new wavelength is different from other wavelengths used in the WDM system. The wavelength conversion enables compatibility between systems and facilitates system expansion.

Techniques for wavelength conversion can be categorized into two types: optoelectronic wavelength conversion and all-optic wavelength conversion. The all-optic wavelength conversion also includes several schemes. They are
- cross gain modulation;
- cross phase modulation;
- four-wave mixing.

5.1 Optoelectronic Wavelength Conversion

Optoelectronic wavelength conversion is the conventional way to achieve this purpose. The optical signals at wavelength $\lambda 1$ are first converted to electronic signals using a receiver, which is the photodetector discussed in Chapter 5. Then an electronic circuit is used to restore and amplify the electronic signals so that they replicate the original signals without distortion.

In a later stage, the electronic signals are converted back to optical signals using a laser diode operating at another wavelength λ2 and are then delivered to the optical fiber. The block diagram of process is shown in Fig.8-47.

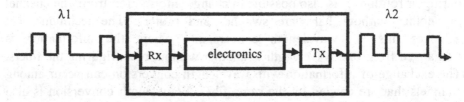

Figure 8-47. Optoelectronic wavelength conversion.

The optoelectronic methods are still popular because all-optic wavelength conversion techniques are not yet mature enough. In addition, the cost to implement the process shown in Fig. 8-47 continues to drop due to the mass production of photodetectors, electronics, and laser diodes.

There are also disadvantages associated with optoelectronic wavelength conversion. The optical signals have to be modulated in a specific format and a specific bit rate. New signal formats will require new electronics for restoration of signals. In addition, the phase information is lost, so the formats of optical signals involving phase modulation, frequency modulation, or analog amplitude variation cannot be used.

5.2 Cross Gain Modulation (XGM)

Cross gain modulation is caused by the mode competition of a gain medium between two modes. The gain competition can be briefly described by the following equation. [23, 24]

$$\frac{dI_1}{dt} = (\frac{g_1}{1 + S_1 I_1 + C_{12} I_2} - l_1) I_1 \qquad (8\text{-}11)$$

where S_1 is the self-saturation coefficient, C_{12} is the cross-saturation coefficient, g_1 is the gain of the laser mode 1, l_1 is the loss that the laser mode 1 experiences, I_1 is the intensity of the laser mode 1, and I_2 is the intensity of a different laser mode that coexists with laser mode1 in the same gain medium.

In Chapter 3, we explained gain saturation due to the intensity of the laser mode, called laser mode 2. Eq. (8-11) tells us that the saturation of the gain medium is caused not only by the laser mode 1, but also by another laser mode. Therefore, when the intensity of the laser mode 2 increases, the gain of laser mode 1 decreases, so its intensity decreases. When the intensity of laser mode 2 fluctuates, the intensity of the laser mode 1 varies, too, but with the inverted shape. The function of cross gain modulation is illustrated in Fig. 8-48.

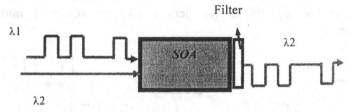

Figure 8-48. Wavelength conversion using cross gain modulation between wavelengths λ1 and λ2.

Semiconductor optical amplifiers (SOAs) are often used for cross gain modulation. When laser beams of two modes are injected into the SOA, they compete for the total carrier density in the quantum wells. When one mode increases the intensity, it depletes the carrier density, causing the gain to reduce. As a result, the intensity of the other mode decreases. The carrier density actually inversely replicates the fluctuation patterns of the laser mode, called pump, resulting in gain variation in a similar manner. The intensity of the other mode, called signal, then varies accordingly. [25] Thus they behave like mode competition in a laser. The converted signals actually have a pulse shape that is the inverse of the input signals, as shown in Fig. 8-48. The wavelength conversion with XGM using SOAs can be operated with the signal bit rate as high as 40 Gb/s. [26]

5.3   Cross Phase Modulation (XPM)

One disadvantage of using SOAs for XGM is that the variation of carrier density causes the refractive index to change. It then further leads to phase modulation, resulting in the chirp of the converted signal like the direct modulation of laser diodes described in Chapter 3. On the other hand, the change of the refractive index can be utilized as an advantage. The change of the refractive index leads to the phase shift. When the SOAs are used in the two paths of a Mach-Zehnder interferometer, the phase shift caused by the laser beam creates the interference for cross phase modulation

Fig. 4-49 shows the configuration of using SOAs for cross phase modulation. The light at wavelength λc is incident into the Mach-Zehnder interferometer from the left-hand side. This light carries no signals. It is split into two paths by the 3 dB coupler. The light with signals at wavelength λs is incident into the Mach-Zehnder interferometer from the right-hand side. This light is also split into two paths by the 3 dB coupler. The fluctuation of light due to the optical signals at wavelength λs results in the phase modulation in the two paths of the modulator. The phase modulation causes the interference to be constructive or destructive, depending on the phase shift. Thus the signals at wavelength λs are converted into light at wavelength λc.

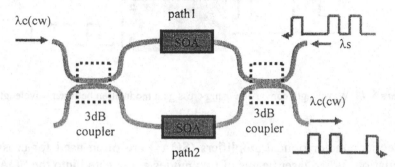

Figure 8-49. Wavelength conversion using cross phase modulation.

The phase shift of π for completely destructive interference can be obtained with only a few dB of the gain suppression in the SOA. This can give a very good extinction ratio. However, the phase shift can not be over π. Otherwise, there will be an interference overshoot that can severely degrade the converted signals. This can be avoided with proper bias conditions on the SOAs or adjusting the intensity of the data signals using preamplifiers. [27]

The converted signals can be in the inverted or non-inverted pulse shape of the input signals, depending on the bias condition. The data signals can also be injected from the same side of the cw beam or the opposite side, as shown in Fig.4-49. The later case (shown in Fig. 4-49) has the advantage of not requiring a filter to separate the original signals and the converted signals.

The wavelength conversion with XPM using SOAs can be operated with the signal bit rate as high as 100 Gb/s with a one-bit-period delay in one of the two paths. [28]

5.4   Four-Wave Mixing (FWM)

Four-wave mixing is a nonlinear phenomenon of optics. As we discussed in Sec. 1.5.6, the polarization induced by the incident light can be nonlinearly proportional to the field strength. In general, the polarization $\bar{P}$ is related to the electrical field $\bar{E}$ according to Eq. (1-47). The third-order susceptibility χ_3 is responsible for the four-wave mixing. The polarization caused by the third-order susceptibility χ_3 is written in the following.

$$\bar{P}_{NL} = \varepsilon_0 \chi_3 \bar{E}\bar{E}\bar{E} \qquad (8\text{-}12)$$

where $\bar{E}$ is the electric field of light, $\bar{P}_{NL}$ is the induced nonlinear polarization. If the light field consists of four frequencies ω_1, ω_2, ω_3, and ω_4, and is x-polarized, the total field will be

$$\bar{E} = \hat{x}\frac{1}{2}\sum_{i=1}^{4} E_i \exp[j(k_i z - \omega_i t) + c.c. \qquad (8\text{-}13)$$

where $k_i = \omega_i n_i /c$. n_i is the refractive index that the light at frequency ω_i experiences in the nonlinear medium. Here the four waves at the four frequencies are assumed to propagate towards the z-direction. Substituting Eq.(8-13) into Eq.(8-12), we obtain the nonlinear polarization that contains the term for four-wave mixing. [29]

$$\bar{P}_{NL} = \hat{x}\frac{3}{4}\{E_1 E_2 E_3^* \exp(j\phi)\exp[j(k_4 z - \omega_4 t)] + c.c. + ...\} \qquad (8\text{-}14)$$

where

$$\phi = (k_1 + k_2 - k_3 - k_4)z - (\omega_1 + \omega_2 - \omega_3 - \omega_4)t \qquad (8\text{-}15)$$

There are more terms in the right hand side of Eq.(8-14). Here we only explicitly express one of the terms that are related to four-wave mixing. The term written in the right hand side of Eq. (8-14) is significant when the phase term in Eq.(8-15) vanishes. The condition is called phase-matching. In this case, we have

$$\omega_1 + \omega_2 = \omega_3 + \omega_4 \qquad (8\text{-}16)$$

If we send three frequencies of light into the nonlinear medium with the third-order susceptibility χ_3, the fourth frequency will be generated according to Eq. (8-16). In SOA, the phenomenon of FWM can be used for wavelength conversion. [10] We do not even have to send three frequencies into the SOA. Two frequencies will be sufficient to cause four-wave mixing. In the SOA, light with two frequencies, say ω_1 and ω_2 and $\omega_1 < \omega_2$, causes the carrier density to oscillate at the beat frequency $f_1 - f_2$, where $\omega_1 = 2\pi f_1$ and $\omega_2 = 2\pi f_2$. As a result, the gain and the refractive index are modulated at the same frequency, creating the side-band frequencies at the two sides of the frequencies ω_1 and ω_2. One of the side-band frequencies is less than ω_1 for the beat frequency. The other side-band frequency is larger than ω_2 for the beat frequency as well. Therefore, the new frequencies are $2\omega_1 - \omega_2$ and $2\omega_2 - \omega_1$. The four-wave mixing of the nonlinear effect further enhances the intensity of light at the new frequencies.

When one of the original frequencies carries optical signals, the light at the new generated frequencies will carry the same optical signals. Fig. 8-50 shows the wavelength conversion using FWM. We usually call the frequency of light with optical signals the "signal frequency" and the other frequency without variation the "pump frequency". Light at two wavelengths λs and λp is sent into the SOA. Due to FWM, two additional wavelengths are generated. The intensity variations at the signal wavelength are transferred to other wavelengths, $2\lambda s - \lambda p$ and $2\lambda p - \lambda s$. The longer wavelength usually has better conversion efficiency and so is usually filtered out as the output signal.

Conversion speed is very fast using FWM. In addition, the phase information of the electric field can be transferred to the new wavelength according to Eq. (8-14). However, the conversion efficiency of FWM is not high due to the weak nonlinear effect, when compared to the usual linear process. This method is not expected to be widely used for wavelength conversion. [30]

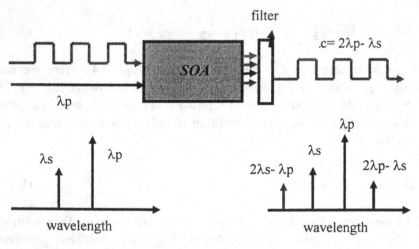

Figure 8-50. Wavelength conversion using four-wave mixing.

REFERENCES

1. Keiser, G., *Optical Fiber Communications*. 2/ed, McGraw-Hill, 1991.
2. Fujii, Y., Minowa, J., and Tanada, H., Practical two-wavelength multiplexer and demultiplexer: design and performance. Applied Opitcs 1983; 22: 3090 – 3097.
3. Senior, J. M., *Optical Fiber Communication-Principles and Practice*. 2/ed, Prentice Hall, 1992.

4. Kashima, Norio, *Passive Optical Components for Optical Fiber Transmission*. Artech House, 1995.
5. Digonett, M and Shaw, H. J., Wavelength multiplexing in single-mode fiber couplers. Applied Optics 1983; 22: 484-491.
6. Smit, M.K. and Dam, C. van, PHASAR-based WDM devices: principles, design, and applications. IEEE Journal of Selected Topics on Quantum Electronics 1996; 2: 236-250.
7. Ishida,O., Takahashi, H., Suzuki, S., and Inoue, Y., Multichannel frequency-selective switching employing an arrayed-waveguide grating multiplexer with fold-back optical paths, IEEE Photonics Technology Letters. 1994; 6: 1219-1221.
8. Hida, Y., Inoue, Y., and Imamura, S., Polymeric arrayed-waveguide grating multiplexeroperating around 1.3 μm. Electronics Letters 1994; 30: 959-960.
9. Smit, M. K., Koonen, T., Herrmann, H., and Sohler, W., "Wavelength-Selective Devices." In *Fiber Optic Communication Devices*. Grote, N. and Venghaus, V., eds. Springer, 2001.
10. Ramaswami, R. and Sivarajan, Optical Networks: A Practical Perspective. Morgan Kaufman, 1998.
11. Lin, L. Y., Goldstein, E. L., Simmons, J. M., Tkach, R. W., High-density connection-symmetric free-space micromachined polygon optical cross-connects with low loss for WDM networks. Proc. OFC'98, PD24-1, San Jose, 1998.
12. Giles, R., Aksyuk, V., Bolle, C., Pardo, F., and Bishoh. D. J., "Silicon Micromachines inOptical Communications Networks: Tiny Machines for Large Systems." In *MEMS and MOEMS Technology and Applications*. Rai-Choudhury, P., ed. SPIE Press, 2000.
13. Su, G.-D. J., Jiang, F., Chiu, E., Avakian, A., Dickson, J., Jia, D., and Tsao, T., Design, test and qualification of stiction-free MEMS optical switches. CLEO/Pacific Rim 2003, Taipei, Taiwan.
14. Agilent Photonic Swiching Platform: N3565A 32 x 32 photonic switch, technical specifications, http://www.agilent .com/
15. Noguchi, K., Optical multichannel switch composed of liquid-crystal light-modulator arrays and bi-refringent crystals. Electronnics Letters 1997; 33: 1627-1629.
16. Crossland, W. A., Manolis, I. G., Redmond, M. M., Tan, K. L., Wilkinson, T. D., Holmes, M. J., Parker, T. R., Chu, H. H., Croucher, J., Handerek, V. A., Warr, S. T., Robertson, B., Bonas, I. G., Franklin, R., Stace, C., White, H. J., Woolley, R. A., and Henshall, G., Holographic optical switching: the 'ROSES' demonstrator. Journal of Lightwave Technology 2000; 18: 1845-1854.
17. Shibata, T., Okuno, M., Goh, T., Yasu, M., Itoh, M., Ishii, M., Hibino, Y., Sugita, A., and Himeno, A., Silica-based 16 x 16 optical matrix switch module with integrated driving circuits, Optical Fiber Communication Conference and Exhibit, 2001. OFC 2001; postdeadline paper, 3: WR1-1 -WR1-3.
18. Rabbering, F. L. W., van Nunen, J. F. P., and Eldada, L., Polymeric 16 x 16 digital optical switch matrix. ECOC 2001; Postdeadline paper, 6: 78-79.
19. Borella, M., S., Jue, J. P., Banerjee, D., Ramamurthym, B., and Mukherjee, B., Optical components for WDM lightwave networks. Proceedings of IEEE 1997; 85: 1274-1307.
20. Al-Salamesh, D. Y., Korotky, S. K., Levy, D. S., Murphy, T. O., Patel, S. H., Richards, G. W., and Tentarelli, E. S., "Optical Switching in Transport Networks: Applications, Requirements, Architectures, Technologies, and Solutions." In *Optical Fiber Telecommunications IVA: Components*. Kaminow, I. and Li, T., eds. Academic Press 2002.
21. Franz J. H. and Jain V. K., *Optical Communications: Components and Systems*. Alpha Science International Ltd., 2000.
22. Hinton, H. S., Photonic switching fabrics. IEEE Communication Magazine 1990; 28: 71-89.
23. Siegman, A. E., *Lasers*. University Press, 1986.

24. Lin, C. F. and Ku, P. C., Analysis of stability in two-mode laser systems. IEEE J. Quantum Electron. 1996; 32: 1377-1382.
25. Joergenson, C., Durhuus, T., Braagaard, C., Mikkelsen, B., and Stubkjaer, K. E., 4 Gb/s optical wavelength conversion using semiconductor optical amplifiers. IEEE Photonics Technology Letters 1993; 5: 657-660.
26. Joergenson, C., Danielsen, S. L., Vaa, M., Mikkelsen, B., Stubkjaer, K. E., Doussiere, P, Pommerau, L. Goldstein, and Goix, M., 40 Gbit/s all-optical wavelength conversion by semiconductor optical amplifiers. Electronics Letters 1996; 32: 367-368.
27. Janz, C., Dagens, B., Bisson, A., Poingt, F., Pommereau, F., Gaborit, F., Guillemot, I., and Renaud, M., Integrated all-active Mach-Zehnder wavelength converter with −10 dBm signal sensitivity and 15 dB dynamic range at 10 Gbit/s. Electronics Letters 1999; 35: 588-590.
28. Leuthold, J., Joyner, C. H., Mikkelson, B., Raybon, G., Pleumeekers, J. L., Miller, B. I., Dreyer, K., and Burrus, C. A., 100 Gbit/s all-optical wavelength conversion with integrated SOA delayed-interference configuration. Electronics Letters 2000; 36: 1129-1130.
29. Agrawal G. P., *Nonlinear Fiber Optics*. 2/ed. Academic Press, 1995.
30. Zhou, J., Park, N., Vahala, K. J., Newkirk, M. A., and Miller, M. I., Four-wave mixing wavelength conversion efficiency in semiconductor traveling-wave amplifiers measured to 65 nm of wavelength shift. IEEE Photonics Technology Letters 1994; 6: 984-987.

INDEX